UNRAVELING THE SAFETY PROFILE OF NANOSCALE PARTICLES AND MATERIALS - FROM BIOMEDICAL TO ENVIRONMENTAL APPLICATIONS

Edited by **Andreia C. Gomes**
and **Marisa P. Sarria**

Unraveling the Safety Profile of Nanoscale Particles and Materials - From Biomedical to Environmental Applications
http://dx.doi.org/10.5772/65837
Edited by Andreia C. Gomes and Marisa P. Sarria

Contributors
Adriele Prina-Mello, Dania Movia, Ramakrishna Podila, Lourdes Rodriguez-Fragoso, Anahi Rodríguez-López, Jorge Reyes-Esparza, Patricia Rodríguez-Fragoso, Gerardo González De La Cruz, Rocío Gómez-Cansino, Subas Chandra Dinda, Nguekam Feugang, Marcelo Grijalva, Javier Camacho, Brajesh Kumar, María José Vallejo, Lizeth Salazar, Nadine Millot, Lionel Maurizi, Anne-Laure Papa, Julien Boudon, Sudhakaran Sruthi, David Vandroux, Gérard Lizard, Johanna Chluba, Benoist Pruvot

Published by InTech
Janeza Trdine 9, 51000 Rijeka, Croatia

Publishing Process Manager Edi Lipovic
Technical Editor SPi Global
Cover InTech Design team

First published March, 2018
Printed in Croatia
Legal deposit, Croatia: National and University Library in Zagreb

Additional hard copies can be obtained from orders@intechopen.com

Unraveling the Safety Profile of Nanoscale Particles and Materials - From Biomedical to Environmental Applications, Edited by Andreia C. Gomes and Marisa P. Sarria
p. cm.
Print ISBN 978-953-51-3939-3
Online ISBN 978-953-51-3940-9

For EU product safety concerns:
IN TECH d.o.o., Prolaz Marije Krucifikse Kozulić 3, 51000 Rijeka, Croatia,
info@intechopen.com or visit our website at intechopen.com.

Meet the editors

Andreia C. Gomes, BSc, PhD, is the vice-director of the Centre of Molecular and Environmental Biology (CBMA) and a member of the Scientific Council of Institute of Science and Innovation for Bio-Sustainability (IB-S). She has been an assistant professor in the Department of Biology at the University of Minho since 2007 and the co-founder of spin-off Nanodelivery-I&D em Bionanotecnologia, Lda.
She has co-authored 3 patents and almost 80 publications and supervised several master's, doctoral and postdoctoral degree students.
Her actual research interests are focused on the study of the interface between nanostructured materials and cells and tissues, so as to optimize biological and/or therapeutic effect with minimal toxicity risk.

Marisa P. Sarria, BSc, PhD, joined the International Iberian Nanotechnology Laboratory (INL), Braga, Portugal, in 2015 as a Marie Curie research fellow. Her COFUND project was devoted to develop innovative nanoagents for delivery of marine toxins towards therapeutic applications. Dr. Sarria's actual research interests focus on investigation of the toxicological profile of nanoparticles and nanomaterials. For the excellence of her academic path, she was granted with three merit awards. In 2011, she was awarded a doctoral degree with distinction in Biology, from the University of Porto, Porto, Portugal. As a postdoctoral researcher at the University of Minho from 2012 to 2014, she collaborated in large-scale European projects committed to develop nanobiodevices for integrated theranostics targeting human diseases.

Contents

Preface

The exponential development of nanotechnological products in recent years has led to pressing concern regarding the safety profiles when considering human and environmental exposure. Nanotoxicology has emerged as a scientific topic of growing interest, and much effort is being put into the development of valid protocols and assays to achieve, as much as possible, the standardization of parameters applied to the validation of nanoparticles and nanomaterials. How their biophysical characteristics influence the crucial interface with living cells and organisms and relevant protocols for this assessment are explored in Section 1 (Biointerfaces with Nanoparticles and Nanomaterials). The following section (Toxicity Validation for Nanomedicine) discusses in detail specific protocols widely used for in vitro testing in cancer research and applications in other fields of nanomedicine. The last section (Environmental Nanotoxicology) approaches the importance of studying bioaccumulation and drawing toxicity profiles of nanoparticles and nanomaterials.

This book aims to give an overview of some of the most important methodologies and developments in nanotoxicity, which will certainly impact the way nanoparticles and other nanostructured materials are perceived and applied in our daily lives.

Andreia C. Gomes
University of Minho
Department of Biology
Campus de Gualtar
Braga, Portugal

Marisa P. Sarria
International Iberian Nanotechnology Laboratory (INL)
Braga, Portugal

Section 1

Biointerfaces with Nanoparticles and Nanomaterials

Chapter 1

Spectroscopic Insights into the Nano-Bio Interface

Achyut J. Raghavendra, Wren Gregory,
Indushekhar Persaud, Jared M. Brown and
Ramakrishna Podila

Additional information is available at the end of the chapter

http://dx.doi.org/10.5772/intechopen.69384

Abstract

Engineered nanomaterials (ENMs) strongly interact with biomolecules due to their unique physicochemical properties. From the standpoint of nanotoxicity, it is imperative to achieve a comprehensive understanding of various nano-bio interactions to ultimately design benign ENMs that do not elicit adverse physiological responses. Spectroscopic tools are ideal for elucidating the underlying biophysical mechanisms of nano-bio interactions. In this chapter, we review spectroscopy techniques, such as Raman, infrared, circular dichroism (CD), and hyperspectral imaging, to illuminate the nano-bio interface. Particularly, we discuss the role of spectroscopic tools in gaining a fundamental understanding of the formation and influence of protein corona on ENM physiological responses.

Keywords: protein corona, spectroscopy, nano-bio interface, silver

1. Introduction

The advancement of nanotechnology over the past two decades has spurred the fields of health care, information technology, energy, homeland security, food safety, and transportation; and the global market for nanotechnology-related products reached more than $200 billion in 2009 with a projected $1 trillion per annum by 2015 (US Senate Committee, 2011) [1]. Despite this enormous global market, there remain several concerns regarding the impact of engineered nanomaterials (ENMs) on biological responses in living organisms and the environment at large [1–7]. A comprehensive knowledge of the ENM-biomolecular interactions is central to applications in nanomedicine, consumer goods, and other unintentional exposures. Currently, there are more than 40 nanopharmaceuticals in routine clinical use, and the patents and publications on nanomedicine have been exponentially increasing [8]. Considering that nanomedicine efforts are a sudden convergence of contrasting scientific disciplines (e.g., materials science, bioengineering, pharmacology), the advancement and acceptance of nanotechnology rely heavily on a holistic interdisciplinary understanding of the impact of fundamental properties of ENMs (such as their morphology, size, defects, and chemical stability) on physiological and environmental systems.

Upon introduction into a biological system, ENMs rapidly associate a variety of macromolecules including proteins, peptides, amino acids, fatty acids, lipids, and other organic matter forming protein biocorona [9–11]. The formation of the corona is dictated not only by the physicochemical properties of the ENM but also by the composition of the physiological environment [12]. The addition of the corona on the surface of the ENM imparts a new distinctive interactive surface, which influences activity, deposition, clearance, and cytotoxicity [13–15]. The biocorona has also been shown to compromise the targeting capacity of functionalized ENMs and subsequently hinder delivery therapy [11, 16, 17]. In addition, the inconsistencies observed between *in vitro* and *in vivo* extrapolation of ENM toxicity are likely contributable to differences in ENM-biocorona formation [9, 18]. Spectroscopic tools are ideal for exploring the biological interactions of ENMs with proteins [19–21]. As discussed in this chapter, combining spectroscopic tools with traditional toxicological studies can provide unique insights into the nano-bio interface that could be used to ultimately design benign ENMs.

2. Spectroscopic insights into protein corona

2.1. Hyperspectral imaging

Gold and silver nanoparticles display a collective and resonant oscillation of surface electrons, known as plasmons, upon light excitation in the visible region (400–700 nm). The surface plasmon resonance is highly sensitive to both nanoparticles aggregation and the dielectric constant of the environment surrounding them. In a biological medium, the presence of protein corona around nanoparticles alters the SPR peak by modifying the dielectric constant [22]. Similarly, their cellular uptake could lead to aggregation, which is known to red shift the SPR peak [23]. Hyperspectral imaging is an excellent tool that combines microscopy and spectroscopy in real time by accumulating reflectance spectrum for each pixel in a micrograph. Thus, the state of nanoparticles and protein corona upon cellular uptake can be gleaned from the hyperspectral micrographs [24–26]. Recently, we explored the cellular uptake of silver nanoparticles (Ag NPs) with and without protein corona using hyperspectral imaging [23]. Our studies revealed intracellular modifications resulting from protein corona formation as shown in **Figure 1**. Changes in the microenvironment of AgNPs were evidently reflected in the shift of plasmon energies allowing us to differentiate between intra- and extra-cellular nanoparticles. Hyperspectral imaging presents an alternative to traditional electron microscopy methods for the identification of nanoparticles and protein corona inside the cell [27–30]. The exhaustive sample preparation needed for electron microscopy, such as encapsulation in a polymer followed by microtoming, often leads to artifacts. For example, the large agglomerates formed during the electron microscopy sample preparation are indistinguishable from agglomerates resulting from the loss of nanoparticles surface coating upon entering biological media.

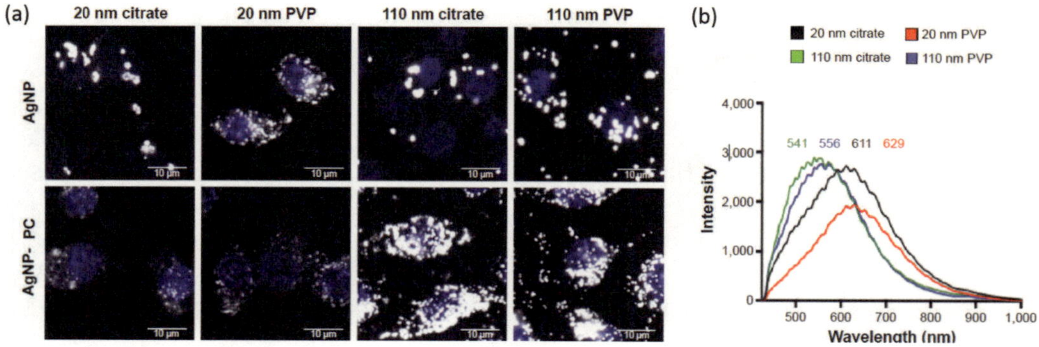

Figure 1. (a) Hyperspectral-enhanced darkfield images of macrophages exposed to AgNPs and AgNPs with protein corona for 2 hours, at a concentration of 25 μg/mL. Macrophage nucleus appears blue in the images due to DAPI stain. (b) Differences in mean spectra for AgNPs with different sizes (20 and 110 nm) and coatings (PVP and citrate). Based on such shifts in plasmon energies, it is possible to study the uptake and modification of AgNPs using hyperspectral imaging.

2.2. Circular dichroism (CD) spectroscopy

ENM-protein complex formation poses concerns on potential denaturation of proteins, which can alter protein binding to receptors and induce inflammatory responses. Lysozyme, for example, adsorbed on gold nanoparticles (AuNPs) was observed to induce misfolded proteins which likely catalyze the formation of aggregates [31]. Similarly, fibrinogen was found to unfold on the surface of stabilized (negatively charged with polyacrylic acid) AuNPs and bind to integrin receptor (MAC-I) leading to inflammatory response [32]. Circular dichroism (CD) spectroscopy has been extensively used to study conformational changes in biomolecules [33–35]. CD studies on ENM protein corona can elucidate the changes in protein secondary structures ensuing from their adsorption to ENMs surface [36]. For example, apolipoprotein (ApoA-I) protein, which is known to be abundant in the protein corona of metal nanoparticles, exhibited secondary structural changes dependent on the surface coating (**Figure 2**). ApoA-I is the major lipoprotein component of high-density lipoprotein (HDL). It adopts a shape similar to a horseshoe of dimensions 12.5 x 8 x 4 nm with high α-helix content [37–39]. The helices in ApoA-I are predicted to be amphipathic, with the hydrophobic (/hydrophilic) face mediating lipid (/aqueous) interactions. The thermodynamic drive to minimize the aqueous exposure of the hydrophobic residues is one of the major factors in ApoA-I adsorption on AgNPs [40, 41]. We studied the interactions between ApoA-I and 100-nm AgNPs with four different coatings, viz., citrate, polyvinylpyrrolidone (PVP), branched polyethylenimine (bPEI), and lipoic acid [42]. These coatings were chosen to provide both negative (citrate, PVP, lipoic acid) and positive charged surfaces (bPEI) with different affinities for AgNPs. While lipoic acid interacts strongly through Ag-S bonds, other coatings (citrate, PVP, bPEI) are considerably weaker. As shown in **Figure 2**, CD studies showed a significant decrease in α-helical content for all surface coatings with the complete disappearance of α-helices for AgNP-bPEI and AgNP-lipoic acid. From a physiological standpoint, we found a significant increase in the ability of ApoA-I-coated AgNP-bPEI and AgNP-lipoic acid to generate reactive oxygen species due to its unfolding. The conformational changes observed in CD can provide more information on ENM-protein corona complex and may be even used as a predictor for adverse immune responses.

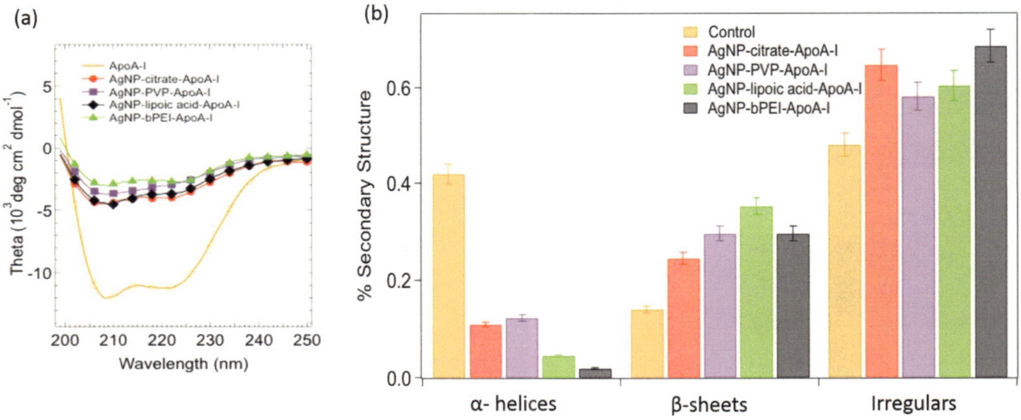

Figure 2. (a) Circular dichroism (CD) spectra for ApoA-I incubated with AgNPs of different surface coatings (citrate, PVP, bPEI, and lipoic acid) show (b) marked decrease in α-helix content with corresponding increase in β-sheets and irregular structures. Spectra were analyzed using CAPITO (a CD Analysis & Plotting Tool), and secondary structure content was estimated. CD spectroscopy provides secondary and tertiary protein structure content estimation of biomolecules, which is critical for analyzing conformational changes during ENM-protein interactions.

2.3. Infrared spectroscopy

Unlike metal nanoparticles, carbon-based ENMs exhibit strong optical absorption <240 nm due to their π-electron system, which can interfere with protein CD spectra collected in 200–300 nm range, precluding the use of CD to study nanocarbon-protein corona. Alternatively, we used ATR-FTIR (attenuated total internal reflection-Fourier transform infrared) spectroscopy to elucidate the adsorption-induced structural changes in proteins (bovine serum albumin or BSA and fibrinogen) on carbon ENMs such as multi-walled carbon nanotubes (MWNTs), graphene, and graphene oxide nanoribbons (GNRs and GONRs). The FTIR spectrum of proteins displays two main bands, Amide I (1600–1700 cm^{-1}) and Amide II (1500–1580 cm^{-1}), arising from the amide bonds that link the amino acids in proteins. While Amide I band is mainly associated with the C=O stretching vibration, Amide II band results primarily from bending vibrations of the N—H bond. Given that both the C=O and the N—H bonds are involved in the hydrogen bonding between the different peptide units, their spectral position and intensity can be used to determine the secondary structure content of a protein. For example, as shown in **Figure 3**, the α-helical content in BSA leads to strong adsorption of ~1640–1660 cm^{-1} (dashed lines), while the lower frequency component at ~1620–1640 cm^{-1} and the peak ~1555 cm^{-1} arise from β-sheets. Clearly, the rich secondary structure of BSA (particularly, the peak relating to α-helical content) significantly disappears upon its adsorption on to all carbon-based NMs, as expected from its low internal stability. Indeed, the changes in secondary structure are higher in the case of MWNTs (i.e., complete disappearance of secondary structure), suggesting that BSA unfolds much more, relative to GNRs and GONRs, in order to adhere to the tubular MWNTs. GNR and GONR retain BSA secondary structure to certain extent, as shown by the presence of ~1555 cm^{-1} for β-sheets. In the case of fibrinogen, the secondary structural changes are found to be higher for GONRs compared to MWNTs and GNRs plausibly due to the formation of hydrogen bonds. The α-helix peak was found to partially disappear for fibrinogen adsorbed on MWNTs and GNRs. Lastly, the

http://dx.doi.org/10.5772/intechopen.69384

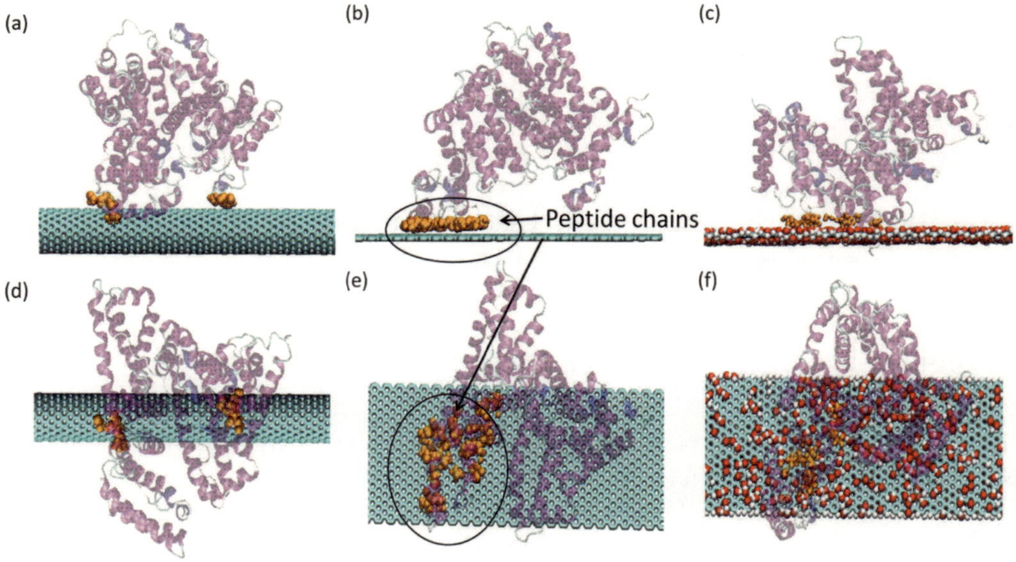

Figure 3. (a–c) Snapshots of MD simulations showing BSA on SWNT, GNR, and GONR 10 ns after adsorption. (d–f) show the top view. The adsorption is accompanied by disruption of α-helices into random peptide chains.

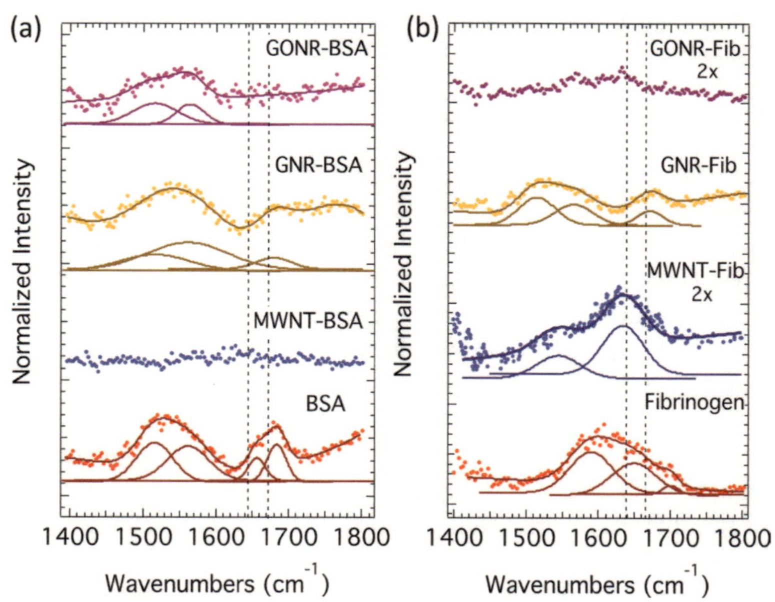

Figure 4. (a) FTIR spectra for native BSA and BSA adsorbed on MWNTs, GNRs, and GONRs show that the CNMs significantly affected the secondary structures of the proteins. The absorption peak for α-helix in native BSA, in the range from 1640 to 1660 cm^{-1} (indicated by the vertical dashed lines), showed a significant reduction (disappeared for MWNTs) for all CNMs, suggesting that the secondary structure of BSA adsorbed on CNMs was less compact. (b) In the case of fibrinogen, the loss of secondary structure was highest for GONRs possibly due to the formation of hydrogen bonds. While both MWNTs and GNRs exhibited a loss of α-helical content, the appearance of new peak ~1500 cm^{-1} (corresponding to random motifs) in GNRs suggested a lesser degree of relaxation for fibrinogen, compared to MWNTs.

structural changes for fibrinogen on GNRs seemed to be less pronounced than MWNTs possibly due to its shape. It could be rationalized that fibrinogen must unfold more to adhere to MWNTs due to their higher curvature than GNRs. To further understand NM-BSA interactions, we also performed large-scale molecular dynamics (MD) simulations of BSA-MWNT, GNR/GONR-water systems (detailed in [43]). As shown in **Figure 3**, we observed that proteins undergo conformational changes after initial contact (in accordance with our experimental observations, discussed in **Figure 4**), leading to some protein regions collapsing on to the ENM surface rather than gradually spreading from the initial region of contact.

3. Probing electronic interactions at the nano-bio interface

Charge transfer is known to play an important role in several many physiological processes including blood clotting, vitamin absorption, and oxidative stress [44–46]. Differences in electronegativity between proteins and the ENM surface may induce charge transfer which, along with other perturbations in the media (e.g., pH, thermodynamic fluctuations), plays a critical role in protein denaturation, protein-protein interactions, and alterations to the cellular and extracellular redox status. Such interactions are not necessarily specific, with conformational changes possibly resulting in newly exposed charged regions and hydrophobic domains attracting or repelling other surrounding proteins or nanostructures. Electronic properties of ENMs, including their band structure and density of states, have been shown to regulate protein adsorption dynamics via charge transfer, which are evident as shifts in absorption and emission spectra [47, 48].

3.1. Charge transfer during corona formation

The chemisorption of proteins on bulk material surfaces has been known to occur through charge transfer processes. For example, the aromatic amino acids present in BSA (tryptophan, phenylalanine, histidine, and tyrosine) could interact with the unhybridized p_z orbitals of the carbon-based ENMs, via providing a weak acceptor level in the electronic density of states (DOS) to allow partial charge transfer. It may be expected that a surface facilitating higher charge transfer at the nanoscale may lead to stronger surface-protein interactions and a subsequent increase in protein adsorption. Spectroscopic and electrochemical techniques such as micro-Raman, CD, and cyclic voltammetry can be used to analyze and elucidate the influence of charge transfer on protein affinity for ENMs and the alterations in secondary and tertiary structures that occur with adsorption. Among variety of ENMs, nanocarbons possess strong affinity for proteins through hydrophobic and aromatic π-π stacking interactions [49]. Previously, we elucidated the charge transfer interactions between nanocarbons and proteins using micro-Raman spectroscopy. Our results show that strong interaction of proteins (albumin and fibrinogen) with nanocarbons is strongly influenced by charge transfer between them, inducing protein unfolding which enhances conformational entropy and higher protein adsorption. For instance, the UV-visible absorption spectrum of single-wall carbon nanotubes (SWNTs) coated with BSA were found to blue shift significantly, while BSA-coated graphene sheets (both exfoliated and synthesized via chemical vapor deposition) exhibited no such changes [50]. Additionally, micro-Raman spectroscopy revealed alterations in the structure of the G-band (or graphitic band) of BSA-coated SWNTs. The G-band, which is highly sensitive to charge transfer, was found to be upshifted with a Breit-Wigner-Fano line-shape resulting from electron transfer between BSA and SWNTs. No changes were observed for graphene upon protein coating, indicating that ENMs-BSA charge transfer is unique to SWNTs. Concomitantly, FTIR spectroscopy revealed subsequent conformational changes in BSA (**Figure 4**) in the form of uncompacting of α-helices, suggesting hard corona formation on SWNTs as opposed to graphene. These results insinuate that disruption in electrostatics due to ENM-protein charge transfer leads to the breaking of peripheral H-bonds in the α-helices and permanent denaturation of BSA on SWNTs [50, 51].

3.2. Two-dimensional (2D) materials

Unique properties of two-dimensional (2D) nanomaterials, including boron nitride (BN), graphene, and graphene oxide, relate to their electronic interactions with biomolecules. The 2D structure of these ENMs gives rise to π-electron clouds which can interact strongly with other π-electrons in proteins or amino acids [52]. While we predominantly focused on proteins, it is important to note that individual amino acid interactions with ENMs are equally significant indicators of corona formation and cytotoxicity modeling. In particular, aromatic amino acids offer π-electrons similar to 2D ENMs, and therefore exhibit this strong affinity for certain 2D materials through this π-π electron stacking. We studied interactions between aromatic amino acids (tryptophan, tyrosine, and phenylalanine) and 2D nanomaterials (graphene, graphene oxide or GO, and BN) using micro-Raman and photoluminescence (PL) spectroscopy combined with electrochemical characterization. Perturbation in the electronic structure was evident through changes in Raman spectra as shown in **Figure 5**, and results were quantified using CV measurements (**Figure 6**). Downshift in characteristic G and 2D bands in the spectra

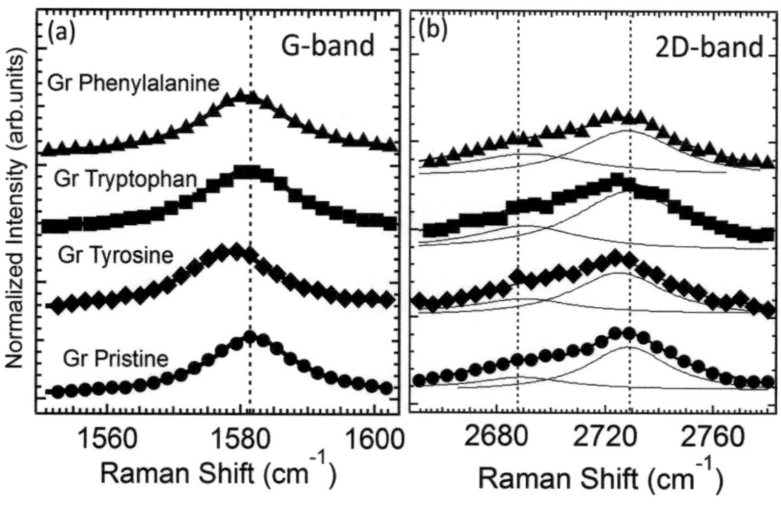

Figure 5. Raman spectra of graphene (Gr)-amino acid complexes show significant shifts in carbon characteristic (a) G band and (b) 2D band indicating charge transfer between amino acid and graphene. Solid line indicates fits to experimental data. Panel b shows deconvoluted peaks of 2D band. Downshift in G and 2D bands for graphene-tyrosine complex indicates upshift in fermi level (E_f) induced by electron transfer from tyrosine.

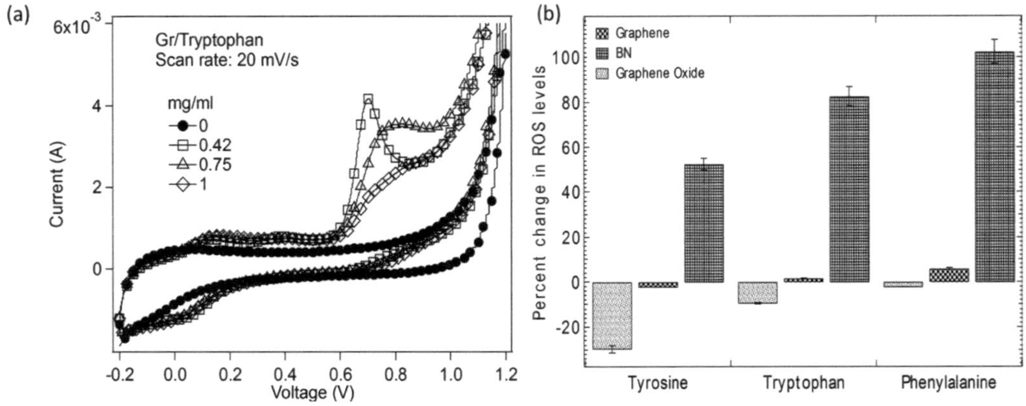

Figure 6. (a) Cyclic voltammetry (CV) curves for graphene and tryptophan show appearance of new peak ~0.6–0.8 V indicating irreversible charge transfer. (b) Percent change in reactive oxygen species (ROS) by macrophages when exposed different 2D materials like graphene (Gr), boron nitride (BN), and graphene oxide (GO) to amino acid complexes (tyrosine, tryptophan, and phenylalanine). Sensitivity to charge transfer makes electrochemical impedance spectroscopy a valuable tool in probing nano-bio interactions.

indicated upshift in Fermi level induced from electron transfer from the amino acid. In CV characterization, the application of gate voltage on the working electrode (i.e., ENMs) modulates its electronic energy levels, which when above (/below) the LUMO (/HOMO) levels of the protein can result in a charge transfer. While reversible charge transfer in redox couples appears as two peaks in the CV curve (one for oxidation and the other for reduction), only a single peak is often observed for irreversible charge transfer. The irreversibility of these charge transfer between 2D ENMs and amino acids was confirmed with appearance of new peak (~0.6–0.8 V) in our CV curves. Further findings suggested variance in π-electron cloud structure in graphene and BN determined electron stacking orientation and associated amino acid affinity, while functional groups mediated affinity through H-bonding in GO. As shown in **Figure 6b**, indeed adsorption of amino acids on 2D materials significantly alters their ability to generate reactive oxygen species [52].

4. Conclusions

Understanding the biological interactions of ENMs with a biocorona and its influence on cellular uptake, generation of reactive oxygen species, and cytotoxicity is critical to implement the safe use of ENMs. Spectroscopic tools offer critical insights into the nano-bio interface by providing information regarding the protein adsorption affinity, the influence of charge transfer, and protein unfolding on ENMs. Hyperspectral imaging is an excellent complementary technique to electron microscopy. The changes in the micro-environment of ENMs are reflected in their hyperspectra allowing one to identify intra-cellular ENMs. Circular dichroism and infrared spectroscopy are sensitive to the changes in protein structure, while Raman and cyclic voltammetry provide information about the charge transfer between ENMs and proteins. A comprehensive characterization of ENM-protein corona with spectroscopic tools is necessary for establishing relationships between ENM physicochemical properties and their biological responses.

Author details

Achyut J. Raghavendra[1,3], Wren Gregory[1,3], Indushekhar Persaud[4], Jared M. Brown[4] and Ramakrishna Podila[1,2,3]*

*Address all correspondence to: rpodila@g.clemson.edu

1 Department of Physics and Astronomy, Clemson Nanomaterials Center, Clemson University, Clemson, SC, USA

2 Center for Optical Materials Science and Engineering Technologies, Clemson University, Clemson, SC, USA

3 Laboratory of Nano-bio Physics, Clemson University, Clemson, SC, USA

4 Department of Pharmaceutical Sciences, the University of Colorado Anschutz Medical Campus, Aurora, CO, USA

References

[1] Wiesner MR, Lowry GV, Alvarez P, Dionysiou D, Biswas P. Assessing the risks of manufactured nanomaterials. Environmental Science and Technology. 2006;**40**:4336-4345

[2] Dror I, Yaron B, Berkowitz B. Abiotic soil changes induced by engineered nanomaterials: A critical review. Journal of Contaminant Hydrology. 2015;**181**:3-16

[3] Colvin V. The potential environmental impact of engineered nanomaterials. Nature Biotechnology. 2003;**21**(10):1166-1170

[4] Farré M, Gajda-Schrantz K, Kantiani L, Barceló D. Ecotoxicity and analysis of nanomaterials in the aquatic environment. Analytical and Bioanalytical Chemistry. 2009;**393**(1): 81-95

[5] Wu YL, Putcha N, Ng KW, Leong DT, Lim CT, Loo SCJ, et al. Biophysical responses upon the interaction of nanomaterials with cellular interfaces. Accounts of Chemical Research. 2013;**46**(3):782-791

[6] Borm P, Castranova V. Toxicology of Nanomaterials: Permanent interactive learning. Particle and Fibre Toxicology. 2009;**6**:28

[7] Fan AM, Alexeeff G. Nanotechnology and nanomaterials: toxicology, risk assessment, and regulations. Journal of Nanoscience and Nanotechnology. 2010;**10**(12):8646-8657

[8] Duncan R, Gaspar R. Nanomedicine(s) under the microscope. Molecular Pharmaceutics. 2011;**8**:2101-2141

[9] Corbo C, Molinaro R, Parodi A, Toledano Furman NE, Salvatore F, Tasciotti E. The impact of nanoparticle protein corona on cytotoxicity, immunotoxicity and target drug delivery. Nanomedicine [Internet]. 2016;**11**(1):81-100. Available from: http://www.futuremedicine.com/doi/10.2217/nnm.15.188

[10] Salvati A, Åberg C, Dawson KA, Monopoli MP, Åberg C, Salvati A, et al. Biomolecular coronas provide the biological identity of nanosized materials. Nature Nanotechnology. 2012;**7**(12):779-786

[11] Tenzer S, Docter D, Kuharev J, Musyanovych A, Fetz V, Hecht R, et al. Rapid formation of plasma protein corona critically affects nanoparticle pathophysiology. Nature Nanotechnology [Internet]. 2013;**8**(10):772-781. Available from: http://www.ncbi.nlm.nih.gov/pubmed/24056901

[12] Shannahan JH, Fritz KS, Raghavendra AJ, Podila R, Persaud I, Brown JM. From the cover: Disease-Induced disparities in formation of the nanoparticle-biocorona and the toxicological consequences. Toxicological Sciences [Internet]. 2016;**152**(2):406-416. Available from: http://www.toxsci.oxfordjournals.org/lookup/doi/10.1093/toxsci/kfw097

[13] Huang J, Bu L, Xie J, Chen K, Cheng Z, Li X, et al. Effects of nanoparticle size on cellular uptake and liver MRI with polyvinylpyrrolidone-coated iron oxide nanoparticles. American Chemical Society Nano. 2010;**4**(12):7151-7160

[14] Yameen B, Choi W Il, Vilos C, Swami A, Shi J, Farokhzad OC. Insight into nanoparticle cellular uptake and intracellular targeting. Journal of Controlled Release. 2014;**190**:485-499

[15] Jiang W, Kim BYS, Rutka JT, Chan WCW. Nanoparticle-mediated cellular response is size-dependent. Nature Nanotechnology [Internet]. 2008;**3**(3):145-150. Available from: http://www.ncbi.nlm.nih.gov/pubmed/18654486

[16] Mirshafiee V, Mahmoudi M, Lou K, Cheng J, Kraft ML. Protein corona significantly reduces active targeting yield. Chemical Communications (Cambridge) [Internet]. 2013;**49**(25):2557-2559. Available from: http://www.pubmedcentral.nih.gov/articlerender.fcgi?artid=3671387&tool=pmcentrez&rendertype=abstract

[17] Arvizo RR, Giri K, Moyano D, Miranda OR, Madden B, McCormick DJ, et al. Identifying new therapeutic targets via modulation of protein corona formation by engineered nanoparticles. PLoS One. 2012;**7**(3):e33650

[18] Podila R, Brown JM. Toxicity of engineered nanomaterials: A Physicochemical Perspective. Journal of Biochemical and Molecular Toxicology. 2013;**27**(1):50-55

[19] Shang L, Nienhaus K, Nienhaus GU. Engineered nanoparticles interacting with cells: Size matters. Journal of Nanobiotechnology. 2014;**12**(1):5

[20] Lundqvist M, Stigler J, Elia G, Lynch I, Cedervall T, Dawson KA. Nanoparticle size and surface properties determine the protein corona with possible implications for biological impacts. Proceedings of the National Academy of Sciences of the United States of America. 2008;**105**(38):14265-14270

[21] Lynch I, Salvati A, Dawson KA. Protein-nanoparticle interactions: What does the cell see? Nature Nanotechnology. 2009;**4**:546-547

[22] Podila R, Chen R, Ke PC, Brown JM, Rao AM. Effects of surface functional groups on the formation of nanoparticle-protein corona. Applied Physics Letters. 2012;**101**(26):263701

[23] Shannahan JH, Podila R, Brown JM. A hyperspectral and toxicological analysis of protein corona impact on silver nanoparticle properties, intracellular modifications, and macrophage activation. International Journal of Nanomedicine. Dove Medical Press Ltd. 2015;**10**:6509-6520

[24] Calin MA, Parasca SV, Savastru D, Manea D. Hyperspectral imaging in the medical field: Present and future. Applied Spectroscopy Reviews. 2014;**49**(6):435-447

[25] Mortimer M, Gogos A, Bartolomé N, Kahru A, Bucheli TD, Slaveykova VI. Potential of hyperspectral imaging microscopy for semi-quantitative analysis of nanoparticle uptake by Protozoa. Environmental Science & Technology. 2014;**48**(15):8760-8767

[26] Fairbairn N, Fernandes R, Carter R, Elliot TJ, Kanaras AG, Muskens OL. Single-nanoparticle detection and spectroscopy in cells using a hyperspectral darkfield imaging technique. Society of Photographic Instrumentation Engineers. 2013;**1**(1):859501

[27] Chen Y, Preece JA, Palmer RE. Processing and characterization of gold nanoparticles for use in plasmon probe spectroscopy and microscopy of biosystems. Annals of the New York Academy of Sciences. 2008;**1130**:201-206

[28] Vetten MA, Tlotleng N, Tanner Rascher D, Skepu A, Keter FK, Boodhia K, et al. Label-free in vitro toxicity and uptake assessment of citrate stabilised gold nanoparticles in three cell lines. Particle and Fibre Toxicology [Internet]. 2013;**10**(1):50. Available from: http://www.pubmedcentral.nih.gov/articlerender.fcgi?artid=3853235&tool=pmcentrez&rendertype=abstract

[29] Patskovsky S, Bergeron E, Meunier M. Hyperspectral darkfield microscopy of PEGylated gold nanoparticles targeting CD44-expressing cancer cells. Journal of Biophotonics. 2015;**8**(1-2):162-167

[30] Grabinski C, Schlager J, Hussain S. Hyperspectral microscopy for characterization of gold nanoparticles in biological media and cells for toxicity assessment. Methods in Molecular Biology [Internet]. 2013;**1025**:167-178. Available from: http://www.ncbi.nlm.nih.gov/pubmed/23918337

[31] Zhang D, Neumann O, Wang H, Yuwono VM, Barhoumi A, Perham M, et al. Gold nanoparticles can induce the formation of protein-based aggregates at physiological pH. Nano Letters. 2009;**9**(2):666-671

[32] Deng ZJ, Liang M, Monteiro M, Toth I, Minchin RF. Nanoparticle-induced unfolding of fibrinogen promotes Mac-1 receptor activation and inflammation. Nature Nanotechnology [Internet]. 2011;**6**(1):39-44. Available from: <Go to ISI>://WOS:000285574000011\nhttp://dx.doi.org/10.1038/nnano.2010.250

[33] Greenfield NJ. Using circular dichroism spectra to estimate protein secondary structure. Nature Protocols [Internet]. 2006;**1**(6):2876-2890. Available from: http://www.nature.com/nprot/journal/v1/n6/abs/nprot.2006.202.html\nhttp://www.nature.com/doifinder/10.1038/nprot.2006.202

[34] Whitmore L, Wallace BA. Protein secondary structure analyses from circular dichroism spectroscopy: Methods and reference databases. Biopolymers. 2008;**89**(5):392-400

[35] Kelly SM, Jess TJ, Price NC. How to study proteins by circular dichroism. Biochimica et Biophysica Acta – Proteins and Proteomics. 2005;**1751**:119-139

[36] Johnson WC Jr. Secondary structure of proteins through circular dichroism spectroscopy. Annual Review of Biophysica and Biophysical Chemistry. 1988;**17**(Cd):145-166

[37] Marcel YL, Kiss RS. Structure-function relationships of apolipoprotein A-I: A flexible protein with dynamic lipid associations. Current Opinion in Lipidology. 2003;**14**(2):151-157

[38] Huang R, Silva RAGD, Jerome WG, Kontush A, Chapman MJ, Curtiss LK, et al. Apolipoprotein A-I structural organization in high-density lipoproteins isolated from human plasma. Nature Structural & Molecular Biology [Internet]. Nature Publishing Group. 2011;**18**(4):416-422. Available from: http://www.nature.com/nsmb/journal/v18/n4/full/nsmb.2028.html?annotation=1

[39] Borhani DW, Rogers DP, Engler JA, Brouillette CG. Crystal structure of truncated human apolipoprotein A-I suggests a lipid-bound conformation. Proceedings of the National Academy of Sciences of the United States of America. 1997;**94**(23):12291-12296

[40] Saha K, Moyano DF, Rotello VM. Protein coronas suppress the hemolytic activity of hydrophilic and hydrophobic nanoparticles. Materials Horizons. 2014;**2014**(1):102-105

[41] Lynch I, Dawson KA. Protein-nanoparticle interactions. Nano Today. 2008;**3**:40-47

[42] Raghavendra AJ, Alsaleh N, Brown JM, Podila R. Charge-transfer interactions induce surface dependent conformational changes in apolipoprotein biocorona. Biointerphases. 2017;**12**(2):02D402

[43] Sengupta B, Gregory WE, Zhu J, Dasetty S, Karakaya M, Brown JM, et al. Influence of carbon nanomaterial defects on the formation of protein corona. RSC Advances [Internet]. Royal Society of Chemistry; 2015;**5**(100):82395-82402. Available from: http://dx.doi.org/10.1039/C5RA15007H

[44] Roy SC, Paulose M, Grimes CA. The effect of TiO2 nanotubes in the enhancement of blood clotting for the control of hemorrhage. Biomaterials. 2007;**28**(31):4667-4672

[45] Morrissey JH, Tajkhorshid E, Rienstra CM. Nanoscale studies of protein-membrane interactions in blood clotting. Journal of Thrombosis and Haemostasis. 2011;**9**:162-167

[46] Wang L, Pal AK, Isaacs JA, Bello D, Carrier RL. Nanomaterial induction of oxidative stress in lung epithelial cells and macrophages. Journal of Nanoparticle Research. 2014;**16**(9):2591

[47] Wang B, Wu P, Yokel RA, Grulke EA. Influence of surface charge on lysozyme adsorption to ceria nanoparticles. Applied Surface Science. 2012;**258**(14):5332-5341

[48] Bradley K, Gabriel J-CP, Briman M, Star A, Grüner G. Charge transfer from ammonia physisorbed on nanotubes. Physical Review Letters. 2003;**91**(21):218301

[49] Firme CP, Bandaru PR. Toxicity issues in the application of carbon nanotubes to biological systems. Nanomedicine: Nanotechnology, Biology, and Medicine. 2010;**6**:245-256

[50] Podila R, Vedantam P, Ke PC, Brown JM, Rao AM. Evidence for charge-transfer-induced conformational changes in CNT-Protien corona. Journal of Physical Chemistry C. 2012;**116**(41):22098-22103

[51] Barros EB, Filho AGS, Lemos V, Filho JM, Fagan SB, Herbst MH, et al. Charge transfer effects in acid treated single-wall carbon nanotubes. Carbon [Internet]. 2005;**43**(12):2495-2500. Available from: <Go to ISI>://WOS:000232093400007\nhttp://linkinghub.elsevier.com/retrieve/pii/S0008622305002599

[52] Mallineni SSK, Shannahan J, Raghavendra AJ, Rao AM, Brown JM, Podila R. Biomolecular interactions and biological responses of emerging two-dimensional materials and aromatic amino acid complexes. ACS Applied Materials & Interfaces. 2016;**8**(26):16604-16611

Chapter 2

Toxicological Risk Assessment of Emerging Nanomaterials: Cytotoxicity, Cellular Uptake, Effects on Biogenesis and Cell Organelle Activity, Acute Toxicity and Biodistribution of Oxide Nanoparticles

Lionel Maurizi, Anne-Laure Papa, Julien Boudon,
Sruthi Sudhakaran, Benoist Pruvot, David Vandroux,
Johanna Chluba, Gérard Lizard and Nadine Millot

Additional information is available at the end of the chapter

http://dx.doi.org/10.5772/intechopen.71833

Abstract

The lack of toxicological data on nanomaterials makes it difficult to assess the risk related to their exposure, and as a result further investigation is required. This chapter presents the synthesis of controlled oxide nanoparticles followed by the evaluation of their safety profile or toxicity (iron, titanium and zinc oxides). The controlled surface chemistry, dispersion in several media, morphology and surface charge of these nanoparticles are presented (transmission electron microscopy, dynamic light scattering, zeta potential, X-ray photoelectron spectroscopy). Classical cytotoxic and cellular uptake studies on different cancer cell lines from liver, prostate, heart, brain and spinal cord are discussed. The incidence of nanoparticles on biogenesis and activity of cell organelles is also highlighted, as well as their biodistribution in animal models. The acute toxicity on zebrafish embryo model is also presented. Finally, the stress is put on the influence and the necessity of controlling the protein corona, a layer of plasma proteins physically adsorbed at the surface of such nanoparticles as a result of their presence in the bloodstream (or relevant biological fluids).

Keywords: superparamagnetic iron oxide nanoparticles (SPIONs), titanate nanotubes (TiONts), zinc oxide, cytotoxicity, oxidative stress, cell organelle activity, zebrafish, cellular uptake, biodistribution, protein corona

1. Introduction

The development and production of nanomaterials are one of the fastest growing areas of advanced technologies, providing a wide range of novel applications in the electronic, health-care, cosmetic, agronomy, engineering and food industries. The nanotechnology era has increased nanoparticles concentration in the environment, causing continuous human exposure, with both uncontrolled contact by inhalation or through the skin, as well as exposure *via* oral administration or by drug injection.

The lack of toxicological data on nanomaterials makes it difficult to assess the risk due to their exposure. For all these reasons, there is an urgent need to develop rapid, accurate and effective testing strategies to assess the impact of these emerging materials on human health and the environment. Three nanoparticles of growing interest have been selected as key materials in this chapter: (1) **superparamagnetic iron oxide nanoparticles** (SPIONs) that are commonly developed as magnetic resonance imaging (MRI) contrast agents [1], (2) **titanate nanotubes** (TiONts) for their elongated morphology [2] and (3) **ZnO** nanoparticles, known for their potential hazards [3].

The major objective of this study concerns the detailed assessment of oxide nanoparticles toxicity or safety profile, through the development of pertinent bioassays. Cytotoxicity assays to check cellular homeostasis disruption, transmission electron microscopy (TEM) analysis and flow cytometry for particle uptake investigation, cell death evaluation, and the influence of nanoparticles on biogenesis and activity of cell organelles are described. Moreover, eco-toxicological monitoring was performed using zebrafish embryos as a model. Survival and hatching rates, and malformations were determined upon exposure to oxide nanoparticles. Finally, an understanding and factors to control the protein/nanomaterial interactions are further presented. These proteins influence the cellular interactions with the nanoparticles such as adhesion to the plasma membrane or uptake, but also their biodistribution.

2. Synthesis, purification and characterization tools for oxide nanoparticles toxicity control and profiling

To investigate the toxicological risk assessment of nanomaterials, it is necessary to jointly control their morphology, their size distribution, the nature of their interfaces (charges and chemistry) and their colloidal stability in several media. Indeed, many controversies in literature come from the lack of control of one of these parameters. In this part, the emphasis is placed on the synthesis as well as the characterization tools used to fully investigate nanoparticles and to control the parameters influencing their nanotoxicity.

2.1. Synthesis routes of oxide nanoparticles: the case of SPIONs and TiONts

2.1.1. Synthesis of SPIONs by soft chemistry as well as functionalization of their surface

SPIONs were prepared according to a method derived from the classical Massart protocol [4]. Briefly, a 1:2 molar ratio of ferrous and ferric chloride was added to a NaOH solution at 90°C under vigorous mechanical stirring. The product was magnetically settled down and washed three times with 400 mL of 1 M HNO_3. Finally, the particle suspension was centrifuged at 450× g for 1 h to remove the biggest aggregates. The supernatant was dialyzed against an HNO_3 solution (pH 4.0) during 24 h [5].

In order to increase their colloidal stability for biological assays, bare SPIONs were subjected to 3-aminopropyltriethoxysilane (APTES) in an equivalent mass ratio into a 1:1 ethanol/water mixture, the pH of which was decreased to 4.0 by the addition of 1 M HCl prior to the APTES addition. The mixture was submitted to an ultrasonic treatment to afford a good particle dispersion leading to the polysiloxane coverage of individual particles rather than agglomerates. The mixture was then submitted to mechanical stirring during 48 h. Glycerol was then added followed by the evaporation of the ethanol/water mixture under reduced pressure to increase the polysiloxane condensation around SPIONs. Finally, glycerol was removed by acetone addition to the SPION suspension accompanied by a magnetic decantation. SPIONs were finally resuspended into ultrapure water yielding SPIONs-NH_2 and dialyzed 1 week against ultrapure water [6].

The surface of bare SPIONs or SPIONs-NH_2 was then functionalized with polyvinyl alcohol (PVA) or polyethylene glycol (PEG-COOH), respectively. Polymers of different molecular weights (from 2 to 30 kDa) and bearing different chemical groups were used. PEG chains were covalently grafted on the surface of SPIONs (EDC/NHS coupling), while PVA was linked *via* electrostatic interactions [7].

2.1.2. *Synthesis of TiONts and functionalization of their surface*

Titanate nanotubes were prepared by a classical hydrothermal method. Titanium dioxide rutile precursor powders (440 mg) were added to a NaOH aqueous solution (10 mol.L^{-1}, 110 mL) [2]. The mixture was subjected to ultrasound (15 min, 375 W) before being transferred into a sealed Teflon reactor. The temperature was set at 155°C for 36 h and the mixture was stirred by magnetic stirring (120 rpm). The resulting white product was isolated by centrifugation and washed with deionized water until pH 6.0 was reached. Finally, the powder was freeze-dried.

The biocompatibility of TiONts in biological systems has been improved through their surface modification with APTES, PEG or chitosan grafting. Bare TiONts were coated by APTES and PEG with protocols very similar to that used for SPIONs [8]. The method of chitosan (CT) grafting is based on electrostatic interactions between chitosan's amines and nanotubes' hydroxyl groups. Briefly, a mixture of TiONts and CT in a 1:2 molar ratio of TiONts(–OH)/CT(–NH_2) was prepared and the pH was adjusted to 7.0. The suspension was mixed up at 25°C under magnetic stirring during 24 h. The powder was washed several times with deionized water by an ultrafiltration device (300 kDa, regenerated cellulose) [9].

2.2. Purifications of bare nanoparticles and nanohybrids

Purification of nanoparticles is undoubtedly an important and challenging step in order to control both the chemistry of their surface and their dispersion in varied media. All the obtained nanohybrids were purified by ultrafiltration on 30 kDa membranes and/or dialyzed on 3.5 kDa membranes to remove ungrafted stabilizing molecules and remaining salts finally yielding nanoparticle suspensions at the pH and the conductivity of ultrapure water. Finally, a portion of the suspensions was freeze-dried for powder characterizations: X-ray diffraction

(XRD), infrared spectroscopy (IR), X-ray photoelectron spectroscopy (XPS) and thermogravimetric analysis (TGA) [8, 10, 11].

2.3. Importance of high-standard characterization tools for nanotoxicity evaluation and safety profile definition

It is well known that numerous parameters influence the toxicity of nanomaterials. This is the reason why each kind of nanoparticles (bare or coated-SPIONs, TiONts or ZnO) was thoroughly characterized by an array of analytical techniques, which allowed to keep track of the precise morphology, composition, agglomeration and surface chemistry (composition and charge). We ensured the highest standards of characterization of hybrid nanomaterials (**Figure 2**).

The size and morphology of the individual nanoparticles were determined by transmission electron microscopy (TEM) and scanning electron microscopy (SEM). The agglomeration state was investigated *via* dynamic light scattering (DLS) and specific surface area measurements (BET method). The colloidal stability was investigated by UV-visible spectroscopy in several media (e.g., PBS, MEM and albumin solutions): the UV absorbance evolution was recorded over time every 5 min during several hours. The faster the absorbance decreased, the less stable the particles were in suspension [9, 10].

The oxide nanoparticles structure was investigated by X-ray diffraction (XRD), high resolution TEM, selected area diffraction (SAD), Raman and FTIR spectroscopies. The nanoparticles chemistry was also analyzed by inductively coupled plasma atomic emission spectroscopy (ICP-AES) (**Figure 2**) [6].

Moreover, the chemical composition of the nanoparticles surface was investigated through XPS, TGA, zeta potential measurements and FTIR spectroscopy. For instance, XPS was associated with TGA to quantify the molecule number grafted at the surface of nanohybrids [12].

3. Cytotoxicity, cellular uptake and biodistribution of oxide nanoparticles

The previously described metal oxide nanoparticles are excellent candidates for a variety of biomedical applications ranging from drug delivery to diagnostic aids, as well as implantable biomaterials [6, 8, 13, 14]. However, a thorough evaluation of particles interaction with the targeted cells, circulatory immune cells and tissues is the first prerequisite. Here, these interactions are presented from an internalization and cytotoxicity point of view, and the biodistribution of these particles in various *in vivo* models (healthy and tumor bearing rodents) is also highlighted.

3.1. Superparamagnetic iron oxide nanoparticles: toxicity, cellular interactions and biodistribution

3.1.1. Cytotoxicity of SPIONs

SPIONs are well developed for biomedical applications as a consequence of their easy and reproducible production. Another advantage of the SPIONs is their chemistry. They are in fact made from one of the most abundant metals present in metabolism: iron. Despite the fact that iron could induce ROS generation in cells [15], working with this element clearly decreases the potential toxicity compared to other metal oxide nanoparticles during dissolution processes *in vitro* or *in vivo* [15]. Naked SPIONs tend to sediment and can precipitate at physiological conditions, leading to a severe toxicological hazard [16]. It is imperative to modify the surface of these nanoparticles to avoid any aggregation and the resultant risk of toxicity. Polyethylene glycol (PEG) is a polymer commonly used to increase the biocompatibility of the SPIONs as well as their stealthiness for specific targeting [17]. SPIONs grafted with PEG do not show any cytotoxicity *via* MTT assay at a concentration up to 270 μgFe/mL, 24 h after incubation with RAW 264.7 and HepG2 cell lines [5]. Polyvinyl alcohol (PVA) is also used to stabilize SPIONs and do not have any cytotoxic effects (MTS assay) for a concentration of 800 μgFe/mL 24 h after incubation with RAW 264.7 cells [18]. Furthermore, PVA coated SPIONs do not significantly activate or influence human T helper cells and have a negligible influence on T cell apoptosis at a concentration of 100 μgSPIONs/mL after 72 h [19]. Regarding PVA, the covalent binding of this polymer onto the SPION surface significantly decreases some inflammatory effects on same T helper cells [20]. Evaluating cytotoxicity of SPIONs is not a trivial operation. Many tests are measuring the evolution of absorbance and the nanomaterials are influencing the final value. It is then very important in order to avoid false positive or negative results, to carefully setup the experiments and the control to correct the absorbance [21].

As demonstrated in many studies, the addition of a biocompatible polymer layer on the SPION surface significantly improves their biocompatibility, which is a crucial step for the biological interactions targeted.

3.1.2. Cellular uptake and biodistribution of superparamagnetic iron oxide nanoparticles

First of all, the magnetic properties of SPIONs are very interesting to increase the cellular uptake rate of these nanoparticles with a magnet [22] and then to improve the labeling of cells for biomedical applications [23, 24]. The concentration and the charge play a significant role in the cellular internalization [25]. For instance, negatively charged fluorescently labeled SPIONs have a higher internalization in prostate-cancer PC-3 cell line as observed *via* confocal microscopy or flow cytometry, in comparison to positively charged SPIONs [26]. In the same way, neutrally charged SPIONs coated with PEG show less cellular interactions on RAW 264.7 cells by TEM and classical microscopy (labeled with Prussian blue) than negatively charged PEGylated-SPIONs [5]. On another side, positively charged PVA-SPIONs seem to have a better internalization by HeLa cells than neutral PVA-SPIONs with an additional influence of the culture medium used, especially depending on the presence of proteins [7]. Thus, the influence of the chemical coating is an important factor, however it seems that the nature of the medium used is much more critical.

Regarding their biodistribution, SPIONs usually show accumulation in the liver and spleen [1, 27]. SPION accumulation to the liver is delayed due to a coating of neutral PEG onto SPIONs, as observed by MRI and Prussian blue-based histology [5]. Their circulation time in the bloodstream is at least 3 h longer compared to negatively charged PEG-SPIONs [5]. The charge of PVA conjugated particles can also induce different *in vivo* behavior of SPIONs. When observed 15 min after injection into a rat model, 50% of the dose of positively charged PVA-SPIONs already accumulate in the liver when 90% of the neutral and the negatively charged PVA-SPIONs are still passing through the bloodstream [28].

Overall, SPIONs are well developed and tested for many biological applications. They do not show any dramatic toxicity and have interesting cellular and *in vivo* interactions, making them extremely attractive as theranostic agents [29].

3.2. Titanate nanotube *in vitro* toxicity and biodistribution testing

3.2.1. Cytotoxicity of TiONts: the surface chemistry matters

The cytotoxicity of titanate nanotubes made by hydrothermal treatment has been assessed in H596 human lung tumor cells [30], cardiomyocytes [31], SNB19 and U87-MG glioblastomas [32], Caco-2 cells [33], as well as 22Rv1 prostate cancer cells [8]. Interestingly, the degree of ion exchange via acid treatment, which partly or entirely substitutes sodium cations by hydrogen cations, is a key parameter that drives the cytotoxicity of titanate nanowires [30]. Titanate nanotubes that were not treated with acid did not induce significant cytotoxic effect in cardiomyocytes, as seen by MTT assay performed between 1 and 10 μg/mL TiONts [31]. Similarly, TiONts concentration ranging up to 100 μg/mL do not induce detectable cytotoxicity in glioblastoma cell lines [32]. In contrast, the viability of CHO cells significantly decreases to 66% viability after 24 h of incubation with 100 μg/mL of bare TiONts; however, concentrations up to 20 μg/mL were not found to be cytotoxic to these ovarian cells [10]. Additionally, PEGylation of TiONts did not modulate TiONts effect on cell viability up to 5 μg/mL [10]. However, the subsequent surface functionalization of these nanotubes with Docetaxel reduces the drug availability and significantly increases Docetaxel IC50 in 22Rv1 cells, compared to free drug control [8]. As described in the next section, *in vivo* studies have shown that TiONts acts like an anchor in the tumor, which prevents drug from leaching out of the cancerous cells, and as a result, the loss in drug potency was not detrimental in this specific case [8].

Beyond their cytotoxicity data, TiONts have been shown to arrest cell cycle in the G2/M phase for both SH-B19 and U87-MG cell lines, as observed while investigating the origin of their radio sensitization effect in glioblastoma [32]. Indeed, an important intrinsic feature of these metal oxide nanotubes is their ability to potentiate gamma radiation effect on cells, making them an interesting candidate for combinatorial therapies on ongoing preclinical investigations (i.e., chemotherapy along with radiation therapy) [34].

3.2.2. Cellular uptake of nanotubes: the shape takes the lead

This chapter mainly focuses on metal oxide nanoparticles, however, beyond the surface chemistry of such materials, one of the key parameters to consider while studying nanoparticle interaction with cells and tissues is their shape. Indeed, due to their needle-like morphology, bare TiONts are internalized in cells not only by endocytosis, but also by diffusion across the plasma membrane, as observed by TEM analysis for cardiomyocytes [31], SNB19 and U87-MG glioblastoma cell lines [32]. Nanotubes display a significant higher specific surface compared to their spherical counterparts [2] and this potentially modifies their degree of interaction with plasma proteins and cells. Our group has observed that even by incubating 4 times more spherical TiO_2 than TiONts with cardiomyocyte cells to account for the difference in specific surface values, TiONts were internalized in much more cells than spherical TiO_2 [31]. Cell penetration via diffusion, along with their increased specific area, potentially makes them an excellent candidate as a new nanomedicine platform after careful assessment of their cytotoxicity in each targeted cell model.

3.2.3. Biodistribution and "tumor retention effect" of titanate nanotubes

Nanotubes display unique behavior regarding their interaction with and internalization within cells, as well as distribution to tissues, compared to spherical nanoparticles. Indeed, the shape is a critical parameter governing circulation time and biodistribution for the same material. For example, the circulation time for tubular micelles in mice is 10 times longer than the one of spherical micelles [35]. Titanate nanotubes have been shown to transiently accumulate in the lungs before being quickly eliminated by the bladder more than 24 h following their IV injection in mice [36]. Lung accumulation has also been observed in the case of carbon nanotubes; however, they were still detected 3 months following injection [37]. Interestingly, single walled carbon nanotubes have been demonstrated to be uptaken in the bloodstream by a subset of monocytes that subsequently deliver them to the tumor [38]. Nanoparticles passively accumulate in tumor by enhanced permeability and retention effect (EPR effect), due to the poorly formed vasculature supporting the malignant cells, in combination with reduced clearance secondary to defective lymphatic drainage at site. While passive targeted delivery to tumor is estimated to deliver only a small fraction of the injected dose utilizing spherical nanoparticles, nanotubes are capable of reaching significantly greater accumulation than their spherical counterparts and also display greater surface area that potentially leads to greater effect [39]. Immune cells' active delivery of tubular nanomaterials to the tumor [38], as well as the enhanced retention time of tubular-shaped nanomaterials into tumors [8] are attractive factors in using such particles for drug delivery. Indeed, our group has reported for the first time that prostate tumors retain more than 70% of docetaxel-functionalized titanate nanotubes up to 7 days following intratumoral injection, indirectly bypassing the well-known multidrug resistance effect [8] (**Figure 1-A**). Exploring the tubular shape of nanobiomaterials that can provide a solid **"tumor retention effect"** will be an important step forward in developing the next generations of drug delivery platforms in oncology.

3.3. Importance of the protein corona on biological interactions of nanomaterials

Understanding the *in vitro* and *in vivo* behavior of nanoparticles is one of the main objectives of current studies. It seems too simplistic today to draw conclusions about their behavior without taking into account the environment, especially the proteins present in the systems studied [40]. Nowadays, it is accepted that once nanoparticles are incubated in biological fluids such as blood, they will be covered by proteins [41]. Not only do these proteins interact with the chemical coatings of materials, but they mostly also modulate their biological fate [28, 42].

The nature of the coating, including resulting charge, surface chemistry and particle hydrodynamic size, influences the adsorption of proteins on the surface of nanoparticles: the protein corona [43, 44]. For example, we demonstrated that bare silica beads covered by either a gold or a titanium oxide layer have different preferential binding to proteins [43]. Once incubated 1 h at 37°C with fetal bovine serum (FBS), we showed that:

- naked silica nanoparticles have no interactions with neither plasminogen nor albumin (two of the most abundant proteins present in FBS),

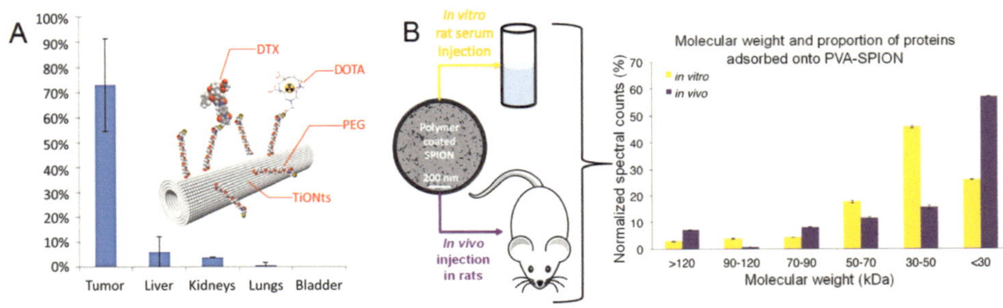

Figure 1. (A) Quantification of nanohybrid biodistribution via gamma counting 7 days post injection (mean of three mice ± SD) (adapted with permission from [8], DTX: docetaxel). (B) Abundance of plasma proteins found at the surface of SPIONs after incubation with rat serum *in vitro* versus *in vivo* (adapted with permission from [28]).

- TiO_2-coated silica nanoparticles interact only with albumin, and
- gold-coated silica nanoparticles interact with both the proteins.

For PVA-coated SPIONs, the charge of the polymers also influences the protein corona [45] and is different for the three types of PVA-SPIONs (neutral, positively and negatively charged) incubated in FBS for 1 h at 37°C. We also showed there were important differences between *in vitro* and *in vivo* protein coronas [28]. In the case of negatively charged PVA-SPIONs, for example, more than 60% of the adsorbed proteins from rat serum have sizes comprised between 30 and 50 kDa *in vitro* when the main proteins (more than 50%) are below 30 kDa *in vivo (*15 min post injection) (**Figure 1-B**). Literature regarding the protein corona of TiONts is very limited at present and our group aims to elucidate key aspects of the topic in the years to come. Interestingly, titanate nanotubes bind significantly less plasma proteins than spherical TiO_2 (Degussa P25) [46], even though they display a greater specific surface [2]. These proteins include albumin, Ig heavy chain (mu), Ig light chain, fibrinogen (alpha, beta and gamma chains) and complement C3.

The coatings of the nanoparticles influence the nature of their protein corona. The medium used is also important for the interactions between materials and proteins [47]. Thus, taking into account not only the physicochemical properties but also the biological environment, it is essential to understand cellular uptake and biodistribution of nanoparticles in order to better control their toxicological risks.

4. Influence of nanoparticles on the biogenesis and activity of cellular organelles

Organelles (mitochondria, peroxisome, lysosome, endoplasmic reticulum, and Golgi apparatus) are integral parts of the cells, essential for the its proper functioning. Their dysfunctions can lead to serious consequences. For instance, mitochondrial alterations can go as far as to activate apoptosis [48], peroxisomal dysfunction affect the mitochondria, subsequently leading to oxidative stress and cell death [49, 50], alterations of the lysosome may have consequences on the induction of autophagy and apoptosis [51], endoplasmic reticulum damages can lead to reticulum stress which can trigger different forms of cell death in extreme cases

[52], and Golgi apparatus dysfunctions can disturb post-translational modifications and vesicular transport [53]. The incidence of the cytotoxicity of nanoparticles is often addressed in generalized terms such as induction of cell death, oxidative stress stimulation, inflammation activation and genotoxicity. The impact of nanoparticles on cell organelles is less known and must be taken into consideration as organelle dysfunctions affect general health in unexpected ways. As regards the peroxisome, whose dysfunctions can lead to severe neurodegenerative damage [54], there are currently no data on the effects of nanoparticles on this organelle.

It is therefore essential to understand the interaction of nanoparticles with cell organelles in terms of distribution and impact on their biogenesis and biological activities. This not only helps to prevent or optimize the toxic effects of nanoparticles depending on the intended purpose (cytoprotection or cell death induction), but also to use them specifically in nanomedicine without side effects.

4.1. Effect of nanoparticles on mitochondria

The interaction of nanoparticles with the mitochondria (as well as the other organelles) must be approached with the consideration that they are either the consequence of targeted interactions with specifically dedicated functionalized nanoparticles, or a random direct or indirect interaction which leads to unwanted side effects. This second aspect must be systematically taken into consideration, and integrated into a cytotoxic screening procedure which will permit to specify the biological activity of nanoparticles at the mitochondrial level. In order to understand the toxicological interactions of nanoparticles on biogenesis and mitochondrial metabolism, it is necessary to specify whether they interact physically with the mitochondria and accumulate at specific locations such as external membrane, mitochondrial space, internal membrane and cristae. In this context, it has been shown that Gadolinium oxide (Gd_2O_3) nanoparticles, which have a range of biomedical uses, induce mitochondrial apoptosis by acting on Bcl-2 and Bax [55]. Similarly, silver nanoparticles impair mitochondrial activity and decrease cell viability [56]. Nanoparticles interact with mitochondria in different manner, based on their physicochemical nature. TiO_2 nanoparticles, which are present in numerous manufactured products, induce loss of mitochondrial transmembrane potential ($\Delta\Psi m$) and an overproduction of superoxide anions in murine microglial BV-2 cells [57]. After exposure with a high concentration of ZnO nanoparticles [58], BV2 cells undergo an increase in mitochondrial transmembrane potential (**Figure 2**). In addition, MTT assays have highlighted that SPIONs and TiONts can also affect mitochondrial integrity depending on their concentrations (especially at high doses) and surface coating [10]. Since numerous types of nanoparticles are able to induce mitochondrial dysfunctions, which can have dramatic consequences on human health after chronic or acute exposures, a systematic evaluation of the impact of nanoparticles on the mitochondria is required.

4.2. Effect of nanoparticles on the peroxisome

Peroxisome has emerged as a key regulator in overall cellular lipid and reactive oxygen species metabolism. In mammals, these organelles have been recognized as important hubs in redox-, lipid-, inflammatory-, and innate immune-signaling networks. Peroxisomal dysfunctions are associated with important brain diseases [54]. To exert its activities, the peroxisome

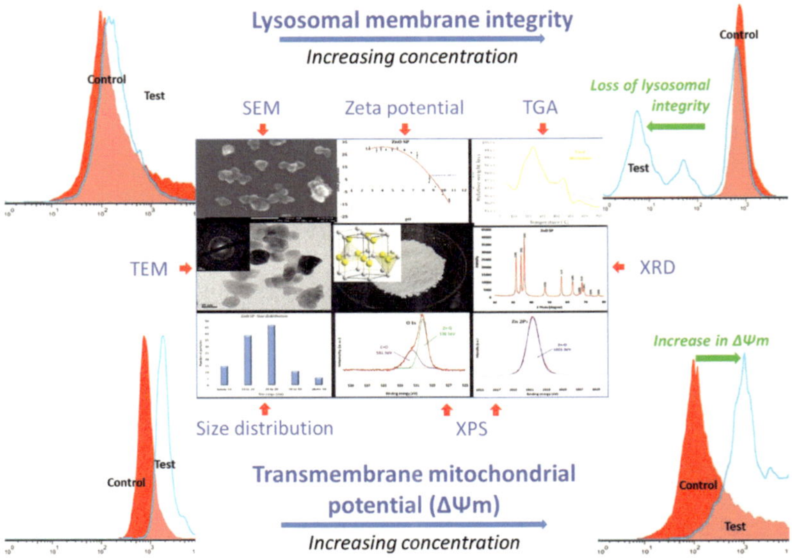

Figure 2. Interaction of ZnO nanoparticles with murine microglial (BV2) cells. Nanoparticles characterized for size, shape, surface charge, crystal structure, chemical composition and purity were exposed to BV2 cells for a maximum duration of 24 h. The ZnO nanoparticles exposure induced dose-dependent increase in transmitochondrial membrane potential and loss of lysosomal membrane integrity as revealed by flow cytometry analysis using fluorescent probes DiOC6(3) and propidium iodide respectively.

must interact both functionally and physically with other cell organelles, mainly mitochondria and endoplasmic reticulum [59, 60]. It seems therefore important to precise the effects of nanoparticles on peroxisome. Numerous techniques such as fluorescent microscopy and flow cytometry are available to estimate the impact of molecules/nanoparticles at the peroxisomal level [61] nevertheless, no data are available concerning the impact of nanoparticles on this organelle.

4.3. Effect of nanoparticles on the lysosome

Endocytosis is the major uptake mechanism of particles by cells [62]. The nanoparticles entrapped in endosomes are eventually degraded by specific enzymes present in phagolysosomes, as the endosomes fuse with lysosomes. The function of lysosomes is to break down molecules and dispose unwanted materials [63]. This phenomenon can also limit the delivery of therapeutic nanoparticles to the intracellular target site. Nanoparticles depending on its physicochemical nature can alter the function of lysosome and subsequently favor the activation or the inhibition of autophagy [64–66]. For instance, we have observed that ZnO nanoparticles induce a loss of lysosome membrane integrity in BV2 cells at high-dose exposure (80 mg/mL) as seen by flow cytometry detection of acridine orange (**Figure 2**). Additionally, double-membrane vesicles closely resembling autophagosomes have been observed by TEM, following 6 h of cardiomyocyte incubation with TiONts [31]. As the lysosomal pathway may have beneficial or detrimental effects on cell activity, a panel of assays is required to define the influence of nanoparticles on this organelle and its potential consequences in major diseases (metabolic diseases, cancer and neurodegenerative diseases).

4.4. Effect of nanoparticles on the endoplasmic reticulum and Golgi apparatus

Currently limited data are available on the impact of nanoparticles on endoplasmic reticulum and Golgi apparatus. It has been reported that silica nanoparticles accumulate in the endoplasmic reticulum and triggers autophagy [67]. On the other hand, the intracellular accumulation of gold nanoparticles leads to inhibition of macropinocytosis and reduction of endoplasmic reticulum stress [68]. Thus, it appears that nanoparticles can have different effects on the endoplasmic reticulum. Consequently, their effects on this organelle must not be neglected.

There is evidence that some nanoparticles can be taken up by the Golgi apparatus for further processing; however, no additional information are available on the influence of nanoparticles on the activity of the Golgi apparatus [69, 70].

Among the most appropriated techniques available in nanotoxicology, observation of cells and tissues by TEM is well suited. This method permits quantitative and qualitative evaluation of modifications at the organelle level which are not easily detected with antigenic and functional changes. Various methods of flow cytometry with appropriate probes are also of interest to define the impact of nanoparticles on the biogenesis and activities of the organelles. These methods can be complemented with other methods of biochemistry, such as Western blot, PCR and RT-qPCR to study the nanobiointeraction at the molecular level. These methods make it possible to identify specific molecular targets and study the effects of nanoparticles on signaling pathways. The development of chip-based single-cell analysis is also of great interest for nanotoxicity assessment [71].

Overall, the beneficial or detrimental effects of nanoparticles on the organelles are difficult to predict. Systemic evaluation of nanoparticle interaction with organelles using simple techniques will help to minimize, if not to subdue, the biological risk associated with nanoparticles on human health, as well as with the environment.

5. Zebrafish as a model for testing the toxicity of SPIONs and TiONts

Due to the increase of nanotechnologies in an expanding range of applications in industrial and biomedical purposes, those new materials require ecotoxicological, biosafety and biocompatibility evaluation. While nanotoxicity can be rapidly assessed *in vitro*, results obtained do not reflect complex processes that happen in full organisms and ecosystems.

Various factors must indeed be taken into account, such as the route of administration (i.e., route of exposure), biodistribution, long-term exposure, induction of developmental defects or activation of the immune system [72, 73]. However, *in vivo* approaches using classical mammalian models have strict ethical considerations, are time consuming and are expensive. Most importantly, throughput approaches cannot be considered *via* those models.

Given its many advantages, zebrafish is now a recognized model for toxicological and biomedical assays [74–77]. The main advantages of this species are rapid development, external fertilization, easy observation of all developmental stages, small size, transparency of the

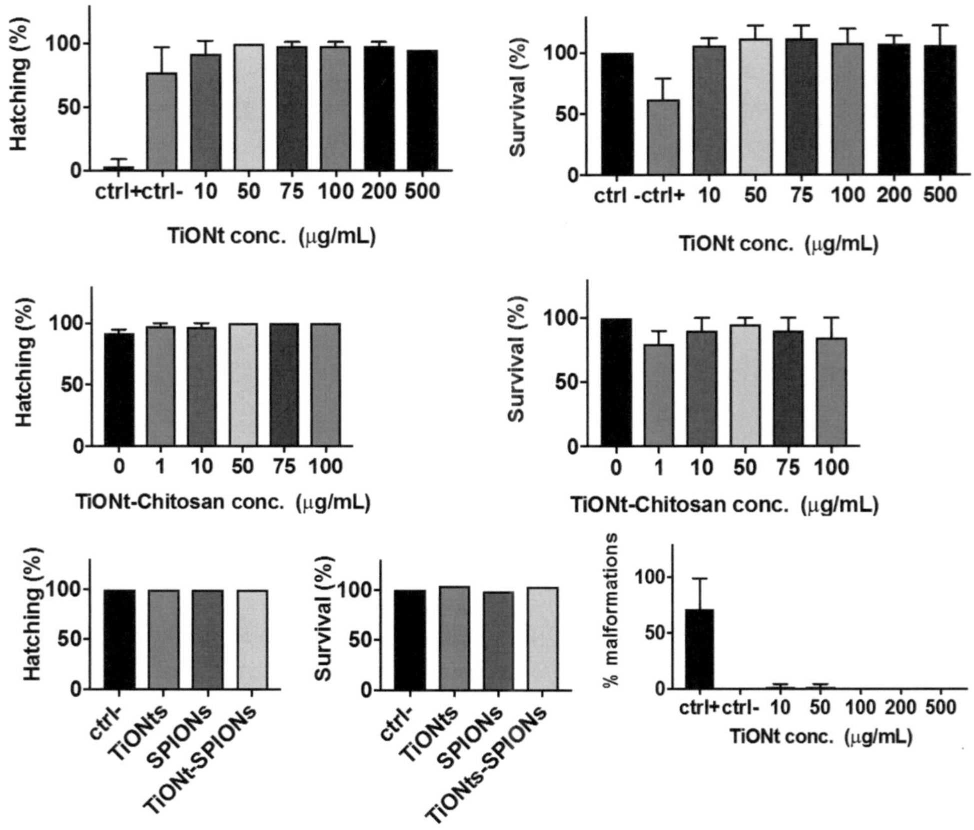

Figure 3. Toxicity evaluation of TiONts and chitosan modified TiONts, SPIONs and TiONt-SPIONs in zebrafish embryos. Unless noted, SPIONs, TiONts and SPION-TiONts concentration was 50 μg/mL; negative control: water; positive control: 4 μg/mL of 3,4-dichloroaniline.

embryos and larvae, large number of embryos, ease of maintenance and close contact with surrounding medium (water) allowing easy interface with materials. In addition, it is possible to take advantage of the sequenced and annotated genome *via* experimental and genetic tools such as fluorescent microscopy, time lapse, histology, transgenic organisms, microinjection and ectopic expression of specific genes. For all these reasons, small fish species represent one of the best choices to study pharmacological/toxicological effects and physiological alterations in vertebrates as a first screening step. Similarly, due to their small size, they are highly suitable for investigating alterations in vertebrate physiology under confined conditions [78]. Indeed, several zebrafish larvae can be placed into one well of 96 wells plates or one larva in a 384 wells plates. Finally, standardized fish embryo toxicity methods are recognized and can be applied to analyze nanomaterials' effects on vertebrates [79].

Recently, zebrafish models were used to evaluate the toxicity of various nanoparticles, including SPIONs and spherical TiO_2 [80–83]. Studies in zebrafish embryos point to toxicity using concentrations of iron oxide particles >10 mg/L resulting in increased mortality, hatching delay and malformations [81], showing the possibility of toxicity of SPIONs at elevated concentrations. Another study revealed that SPIONs coated with cross-linked aminated dextran may cause acute brain toxicity in adult zebrafish [83]. Iron overload, changes of gene expression and inhibition of acetylcholinesterase were proposed as causes for the observed neurotoxicity. In another study, SPIONs, which were non-toxic *in vitro*, were lethal in zebrafish embryos when used at concentrations higher than 10 mg/mL [80]. Regarding TiO_2 nanotubes, they are reported to have an excellent biocompatibility [84]. However, another work using zebrafish as model organism showed that TiO_2 nanotubes at 1 mg/L may cause undesired tissue accumulation in injured animals and asymmetric and shorter regeneration after fin amputation [82]. We analyzed SPIONs as well as unmodified and modified TiONts in zebrafish embryos for toxicity (**Figure 3**). Embryos were incubated up to 96 h post fertilization with different amounts of these materials (500 μg/mL for TiONts, 50 μg/mL for SPIONs and TiONts-SPIONs). No lethality, developmental effects (no malformation) or delayed hatching were observed. Even if we used higher concentrations than Park et al. [82], we did not observe any toxicity. In contrast to Parker, we did not treat injured animals, and we used a higher number of animals in the experiments.

All together, we concluded that the nanomaterials we produced do not show any obvious toxicity, even at high concentrations. However, to get more precise information, the zebrafish embryos will be analyzed by qRT-PCR for detection of stress or inflammation related gene expression and with fluorescent markers for apoptosis events.

6. Conclusion

The outcomes presented in this chapter are the result of collaborations between chemists, physico-chemists, biologists and clinicians on the field of biomedical applications of nanoparticles. Such interdisciplinary collaborations are required to investigate nanotoxicity. Controlled nanoparticles, fully characterized and leading to stable suspensions in biological media, have to be prepared. Then, rapid, accurate and efficient testing strategies have to be developed to assess the effect of these emerging materials on the human health and the environment: *in vitro* assays but also *in vivo* evaluation (biodistribution, retention, elimination and ecotoxicity). All the skills (chemistry, physico-chemistry, nanomaterials engineering, toxicology, biology and medicine) are required to achieve this goal.

Acknowledgements

The authors would like to thank Dr. Vanessa Bellat, Yasmine Saïbi, Thomas Nury, Annette Luce, Fadoua Sallem, Alexis Loiseau, Dr. Guillaume Thomas, Dr. Rémi Chassagnon, colleagues from ICMUB laboratory, from the preclinal imaging platform of the CGFL, from the NVH Medicinal biotechnology company, from the CHU of Dijon. The authors are also indebted to the Université de Bourgogne, Inserm, CNRS, the Raman-Charpak Fellowship and the Conseil Régional de Bourgogne (Contrat d'Etude – CPER 2007–2013 and PARI Nano2Bio), the FEDER program and the Association Française de Cytométrie en Flux.

Author details

Lionel Maurizi[1], Anne-Laure Papa[2], Julien Boudon[1], Sruthi Sudhakaran[3], Benoist Pruvot[4], David Vandroux[5], Johanna Chluba[4], Gérard Lizard[6] and Nadine Millot[1]*

*Address all correspondence to: nadine.millot@u-bourgogne.fr

1 Laboratoire Interdisciplinaire Carnot de Bourgogne, UMR 6303 CNRS/Université de Bourgogne, Dijon, France

2 Department of Biomedical Engineering, School of Engineering and Applied Science, George Washington University, Washington, DC, USA

3 Sree Chitra Tirunal Institute for Medical Sciences and Technology, Kerala, India

4 Lipides Nutrition Cancer, UMR 1231 INSERM/Université de Bourgogne, Dijon, France

5 NVH Medicinal, Dijon, France

6 Bio-PeroxIL EA7270, Université de Bourgogne/INSERM, Dijon, France

References

[1] Mergo PJ, Engelken JD, Helmberger T, Ros PR. MRI in focal liver disease: A comparison of small and ultra-small superparamagnetic iron oxide as hepatic contrast agents. Journal of Magnetic Resonance Imaging. 1998;**8**(5):1073-1078. DOI: 10.1002/jmri.1880080511

[2] Papa A-L, Millot N, Saviot L, Chassagnon R, Heintz O. Effect of reaction parameters on composition and morphology of titanate nanomaterials. Journal of Physical Chemistry C. 2009;**113**(29):12682-12689. DOI: 10.1021/jp903195h

[3] Sruthi S, Mohanan PV. Engineered zinc oxide nanoparticles; biological interactions at the organ level. Current Medicinal Chemistry. 2016;**23**(35):4057-4068

[4] Massart R. Preparation of aqueous magnetic liquids in alkaline and acidic media. IEEE Transactions on Magnetics. 1981;**MAG-17**(2):1247-1248

[5] Maurizi L, Papa A-L, Dumont L, Bouyer F, Walker P, Vandroux D, et al. Influence of surface charge and polymer coating on internalization and biodistribution of polyethylene glycol-modified iron oxide nanoparticles. Journal of Biomedical Nanotechnology. 2015;**11**(1):126-136. DOI: 10.1166/jbn.2015.1996

[6] Boudon J, Paris J, Bernhard Y, Popova E, Decreau RA, Millot N. Magneto-optical nanomaterial: A SPIO-phthalocyanine scaffold built step-by-step towards bimodal imaging. Chemical Communications. 2013;**49**(67):7394-7396. DOI: 10.1039/C3CC41898G

[7] Petri-Fink A, Steitz B, Finka A, Salaklang J, Hofmann H. Effect of cell media on polymer coated superparamagnetic iron oxide nanoparticles (SPIONs): Colloidal stability, cytotoxicity, and cellular uptake studies. European Journal of Pharmaceutics and Biopharmaceutics. 2008;**68**(1):129-137. DOI: 10.1016/j.ejpb.2007.02.024

[8] Loiseau A, Boudon J, Mirjolet C, Créhange G, Millot N. Taxane-grafted metal-oxide nanoparticles as a new theranostic tool against cancer: The promising example of docetaxel-functionalized titanate nanotubes on prostate tumors. Advanced Healthcare Materials. 2017;**6**(16):1700245. DOI: 10.1002/adhm.201700245

[9] Sallem F, Boudon J, Heintz O, Séverin I, Megriche A, Millot N. Synthesis and characterization of chitosan-coated titanate nanotubes: Towards a new safe nanocarrier. Dalton Trans. 2017;**46**:15386-15398

[10] Papa A-L, Boudon J, Bellat V, Loiseau A, Bisht H, Sallem F, et al. Dispersion of titanate nanotubes for nanomedicine: Comparison of PEI and PEG nanohybrids. Dalton Transactions. 2015;**44**(2):739-746. DOI: 10.1039/c4dt02552k

[11] Thomas G, Demoisson F, Chassagnon R, Popova E, Millot N. One-step continuous synthesis of functionalized magnetite nanoflowers. Nanotechnology. 2016;**27**(13):135604. DOI: 10.1088/0957-4484/27/13/135604

[12] Maurizi L, Sallem F, Boudon J, Heintz O, Bisht H, Bouyer F, et al. Surface characterizations to quantify complex one-batch functionalization of iron oxide nanoparticles. Journal of Nanoscience and Nanotechnology. 2017 (Just accepted)

[13] Gupta AK, Gupta M. Synthesis and surface engineering of iron oxide nanoparticles for biomedical applications. Biomaterials. 2005;**26**(18):3995-4021. DOI: 10.1016/j.biomaterials.2004.10.012

[14] Pisanic TR, Blackwell JD, Shubayev VI, Finones RR, Jin S. Nanotoxicity of iron oxide nanoparticle internalization in growing neurons. Biomaterials. 2007;**28**(16):2572-2581. DOI: 10.1016/j.biomaterials.2007.01.043

[15] Sharifi S, Behzadi S, Laurent S, Forrest ML, Stroeve P, Mahmoudi M. Toxicity of nanomaterials. Chemical Society Reviews. 2012;**41**(6):2323-2343. DOI: 10.1039/c1cs15188f

[16] Singh N, Jenkins GJS, Asadi R, Doak SH. Potential toxicity of superparamagnetic iron oxide nanoparticles (SPION). Nano Reviews. 2010;**1**:5358. DOI: 10.3402/nano.v1i0.5358

[17] Lewinski N, Colvin V, Drezek R. Cytotoxicity of nanoparticles. Small. 2008;**4**(1):26-49. DOI: 10.1002/smll.200700595

[18] Maurizi L, Sakulkhu U, Crowe LA, Dao VM, Leclaire N, Vallée J-P, et al. Syntheses of cross-linked polymeric superparamagnetic beads with tunable properties. RSC Advances. 2014;**4**(22):11142-11146. DOI: 10.1039/C3RA48004F

[19] Strehl C, Schellmann S, Maurizi L, Hofmann-Amtenbrink M, Häupl T, Hofmann H, et al. Effects of PVA-coated nanoparticles on human T helper cell activity. Toxicology Letters. 2016;**245**:52-58. DOI: 10.1016/j.toxlet.2016.01.003

[20] Strehl C, Maurizi L, Gaber T, Hoff P, Broschard T, Poole AR, et al. Modification of the surface of superparamagnetic iron oxide nanoparticles to enable their safe application in humans. International Journal of Nanomedicine. 2016;**11**:5883-5896. DOI: 10.2147/IJN.S110579

[21] Bonvin D, Hofmann H, Ebersold MM. Assessment of nanoparticles' safety: Corrected absorbance-based toxicity test. The Analyst. 2017;**142**(13):2338-2342. DOI: 10.1039/C7AN00382J

[22] Petri-Fink A, Hofmann H. Superparamagnetic iron oxide nanoparticles (SPIONs): From synthesis to in vivo studies--a summary of the synthesis, characterization, in vitro, and in vivo investigations of SPIONs with particular focus on surface and colloidal properties. IEEE Transactions on Nanobioscience. 2007;**6**(4):289-297. DOI: 10.1109/TNB.2007.908987

[23] Liang C, Wang C, Liu Z. Stem cell labeling and tracking with nanoparticles. Particle and Particle Systems Characterization. 2013;**30**(12):1006-1017. DOI: 10.1002/ppsc.201300199

[24] Shubayev VI, Pisanic TR, Jin S. Magnetic nanoparticles for theragnostics. Advanced Drug Delivery Reviews. 2009;**61**(6):467-477. DOI: 10.1016/j.addr.2009.03.007

[25] Thorek DLJ, Tsourkas A. Size, charge and concentration dependent uptake of iron oxide particles by non-phagocytic cells. Biomaterials. 2008;**29**(26):3583-3590. DOI: 10.1016/j.biomaterials.2008.05.015

[26] Kralj S, Drofenik M, Makovec D. Controlled surface functionalization of silica-coated magnetic nanoparticles with terminal amino and carboxyl groups. Journal of Nanoparticle Research. 2011;**13**(7):2829-2841. DOI: 10.1007/s11051-010-0171-4

[27] Maurizi L, Sakulkhu U, Gramoun A, Vallee J-P, Hofmann H. A fast and reproducible method to quantify magnetic nanoparticle biodistribution. The Analyst. 2014;**139**(5):1184-1191. DOI: 10.1039/C3AN02153J

[28] Sakulkhu U, Maurizi L, Mahmoudi M, Motazacker M, Vries M, Gramoun A, et al. Ex situ evaluation of the composition of protein corona of intravenously injected superparamagnetic nanoparticles in rats. Nanoscale. 2014;**6**(19):11439-11450. DOI: 10.1039/C4NR02793K

[29] Li K, Nejadnik H, Daldrup-Link HE. Next-generation superparamagnetic iron oxide nanoparticles for cancer theranostics. Drug Discovery Today. 2017;**22**(9):1421-1429. DOI: 10.1016/j.drudis.2017.04.008

[30] Magrez A, Horváth L, Smajda R, Salicio V, Pasquier N, Forró L, et al. Cellular toxicity of TiO_2-based nanofilaments. ACS Nano. 2009;**3**(8):2274-2280. DOI: 10.1021/nn9002067

[31] Papa A-L, Dumont L, Vandroux D, Millot N. Titanate nanotubes: Towards a novel and safer nanovector for cardiomyocytes. Nanotoxicology. 2013;**7**(6):1131-1142. DOI: 10.3109/17435390.2012.710661

[32] Mirjolet C, Papa AL, Créhange G, Raguin O, Seignez C, Paul C, et al. The radiosensitization effect of titanate nanotubes as a new tool in radiation therapy for glioblastoma: A proof-of-concept. Radiotherapy and Oncology. 2013;**108**:136-142. DOI: 10.1016/j.rad onc.2013 .04.004

[33] Fenyvesi F, Kónya Z, Rázga Z, Vecsernyés M, Kása P, Pintye-Hódi K, et al. Investigation of the cytotoxic effects of titanate nanotubes on Caco-2 cells. AAPS PharmSciTech. 2014;**15**(4):858-861. DOI: 10.1208/s12249-014-0115-x

[34] Mirjolet C, Boudon J, Loiseau A, Chevrier S, Boidot R, Oudot A, et al. Docetaxel-titanate nanotubes enhance radiosensitivity in an androgen-independent prostate cancer model. International Journal of Nanomedicine. 2017;**12**:6357-6364. DOI: 10.2147/IJN.S139167

[35] Geng Y, Dalhaimer P, Cai S, Tsai R, Tewari M, Minko T, et al. Shape effects of filaments versus spherical particles in flow and drug delivery. Nature Nanotechnology. 2007; **2**(4):249-255. DOI: 10.1038/nnano.2007.70

[36] Boudon J, Papa A-L, Paris J, Millot N. Titanate nanotubes as a versatile platform for nanomedicine. In: Nanomedicine. One Central Press (OCP); pp. 403-428

[37] Czarny B, Georgin D, Berthon F, Plastow G, Pinault M, Patriarche G, et al. Carbon nanotube translocation to distant organs after pulmonary exposure: Insights from in situ (14) C-radiolabeling and tissue radioimaging. ACS Nano. 2014;**8**(6):5715-5724. DOI: 10.1021/nn500475u

[38] Smith BR, Ghosn EEB, Rallapalli H, Prescher JA, Larson T, Herzenberg LA, et al. Selective uptake of single-walled carbon nanotubes by circulating monocytes for enhanced tumour delivery. Nature Nanotechnology. 2014;**9**(6):481-487. DOI: 10.1038/nnano.2014.62

[39] Toy R, Peiris PM, Ghaghada KB, Karathanasis E. Shaping cancer nanomedicine: The effect of particle shape on the in vivo journey of nanoparticles. Nanomedicine (London, England). 2014;**9**(1):121-134. DOI: 10.2217/nnm.13.191

[40] Forest V, Pourchez J. Preferential binding of positive nanoparticles on cell membranes is due to electrostatic interactions: A too simplistic explanation that does not take into account the nanoparticle protein corona. Materials Science and Engineering C: Materials for Biological Applications. 2017;**70**:889-896. DOI: 10.1016/j.msec.2016.09.016

[41] Lynch I, Dawson KA. Protein-nanoparticle interactions. Nano Today. 2008;**3**(1):40-47. DOI: 10.1016/S1748-0132(08)70014-8

[42] Aoyama M, Hata K, Higashisaka K, Nagano K, Yoshioka Y, Tsutsumi Y. Clusterin in the protein corona plays a key role in the stealth effect of nanoparticles against phagocytes. Biochemical and Biophysical Research Communications. 2016;**480**(4):690-695. DOI: 10.1016/j.bbrc.2016.10.121

[43] Sakulkhu U, Mahmoudi M, Maurizi L, Coullerez G, Hofmann-Amtenbrink M, Vries M, et al. Significance of surface charge and shell material of superparamagnetic iron oxide nanoparticle (SPION) based core/shell nanoparticles on the composition of the protein corona. Biomaterials Science. 2015;**3**(2):265-278. DOI: 10.1039/C4BM00264D

[44] Kurtz-Chalot A, Villiers C, Pourchez J, Boudard D, Martini M, Marche PN, et al. Impact of silica nanoparticle surface chemistry on protein corona formation and consequential interactions with biological cells. Materials Science and Engineering: C. 2017;**75**:16-24. DOI: 10.1016/j.msec.2017.02.028

[45] Sakulkhu U, Mahmoudi M, Maurizi L, Salaklang J, Hofmann H. Protein corona composition of superparamagnetic iron oxide nanoparticles with various physico-chemical properties and coatings. Scientific Reports. 2014;**4**:5020. DOI: 10.1038/srep05020

[46] Deng ZJ, Mortimer G, Schiller T, Musumeci A, Martin D, Minchin RF. Differential plasma protein binding to metal oxide nanoparticles. Nanotechnology. 2009;**20**(45):455101. DOI: 10.1088/0957-4484/20/45/455101

[47] Bonvin D, Aschauer U, Alexander DTL, Chiappe D, Moniatte M, Hofmann H, et al. Protein corona: Impact of lymph versus blood in a complex in vitro environment. Small. 2017;**13**(29):1700409. DOI: 10.1002/smll.201700409

[48] Galluzzi L, Kepp O, Kroemer G. Mitochondrial regulation of cell death: A phylogenetically conserved control. Microbial Cell. 2016;**3**(3):101-108. DOI: 10.15698/mic2016.03.483

[49] Fransen M, Nordgren M, Wang B, Apanasets O. Role of peroxisomes in ROS/RNS-metabolism: Implications for human disease. Biochimica et Biophysica Acta. 2012; **1822**(9):1363-1373. DOI: 10.1016/j.bbadis.2011.12.001

[50] Nordgren M, Fransen M. Peroxisomal metabolism and oxidative stress. Biochimie. 2014; **98**:56-62. DOI: 10.1016/j.biochi.2013.07.026

[51] Kroemer G, Mariño G, Levine B. Autophagy and the integrated stress response. Molecular Cell. 2010;**40**(2):280-293. DOI: 10.1016/j.molcel.2010.09.023

[52] Ferri KF, Kroemer G. Organelle-specific initiation of cell death pathways. Nature Cell Biology. 2001;**3**(11):E255-E263. DOI: 10.1038/ncb1101-e255

[53] Galluzzi L, Bravo-San Pedro JM, Kroemer G. Organelle-specific initiation of cell death. Nature Cell Biology. 2014;**16**(8):728-736. DOI: 10.1038/ncb3005

[54] Trompier D, Vejux A, Zarrouk A, Gondcaille C, Geillon F, Nury T, et al. Brain peroxisomes. Biochimie. 2014;**98**:102-110. DOI: 10.1016/j.biochi.2013.09.009

[55] Alarifi S, Ali H, Alkahtani S, Alessia MS. Regulation of apoptosis through bcl-2/bax proteins expression and DNA damage by nano-sized gadolinium oxide. International Journal of Nanomedicine. 2017;**12**:4541-4551. DOI: 10.2147/IJN.S139326

[56] Braydich-Stolle L, Hussain S, Schlager JJ, Hofmann M-C. In vitro cytotoxicity of nanoparticles in mammalian germline stem cells. Toxicological Sciences. 2005;**88**(2):412-419. DOI: 10.1093/toxsci/kfi256

[57] Rihane N, Nury T, M'rad I, El Mir L, Sakly M, Amara S, et al. Microglial cells (BV-2) internalize titanium dioxide (TiO_2) nanoparticles: Toxicity and cellular responses. Environmental Science and Pollution Research International 2016;**23**(10):9690-9699. DOI: 10.1007/s11356-016-6190-7

[58] Sruthi S, Millot N, Mohanan PV. Zinc oxide nanoparticles mediated cytotoxicity, mitochondrial membrane potential and level of antioxidants in presence of melatonin. International Journal of Biological Macromolecules. 2017;**103**:808-818. DOI: 10.1016/j.ijbiomac.2017.05.088

[59] Lismont C, Nordgren M, Van Veldhoven PP, Fransen M. Redox interplay between mitochondria and peroxisomes. Frontiers in Cell and Development Biology. 2015;**3**:35. DOI: 10.3389/fcell.2015.00035

[60] Fransen M, Lismont C, Walton P. The peroxisome-mitochondria connection: How and why? International Journal of Molecular Sciences. 2017;**18**(6):1126. DOI: 10.3390/ijms18061126

[61] Debbabi M, Nury T, Helali I, Karym EM, Geillon F, Gondcaille C, et al. Flow cytometric analysis of the expression pattern of peroxisomal proteins, Abcd1, Abcd2, and Abcd3 in BV-2 murine microglial cells. Methods in Molecular Biology. 2017;**1595**:257-265. DOI: 10.1007/978-1-4939-6937-1_25

[62] Sakhrani NM, Padh H. Organelle targeting: Third level of drug targeting. Drug Design, Development and Therapy. 2013;**7**:585-599. DOI: 10.2147/DDDT.S45614

[63] Stern ST, Adiseshaiah PP, Crist RM. Autophagy and lysosomal dysfunction as emerging mechanisms of nanomaterial toxicity. Particle and Fibre Toxicology. 2012;**9**:20. DOI: 10.1186/1743-8977-9-20

[64] Song W, Popp L, Yang J, Kumar A, Gangoli VS, Segatori L. The autophagic response to polystyrene nanoparticles is mediated by transcription factor EB and depends on surface charge. Journal of Nanobiotechnology. 2015;**13**:87. DOI: 10.1186/s12951-015-0149-6

[65] Lin Y-X, Gao Y-J, Wang Y, Qiao Z-Y, Fan G, Qiao S-L, et al. pH-sensitive polymeric nanoparticles with gold(I) compound payloads synergistically induce cancer cell death through modulation of autophagy. Molecular Pharmaceutics. 2015;**12**(8):2869-2878. DOI: 10.1021/acs.molpharmaceut.5b00060

[66] Bourdenx M, Daniel J, Genin E, Soria FN, Blanchard-Desce M, Bezard E, et al. Nanoparticles restore lysosomal acidification defects: Implications for Parkinson and other lysosomal-related diseases. Autophagy. 2016;**12**(3):472-483. DOI: 10.1080/15548627.2015.1136769

[67] Wei F, Wang Y, Luo Z, Li Y, Duan Y. New findings of silica nanoparticles induced ER autophagy in human colon cancer cell. Scientific Reports. 2017;**7**:42591. DOI: 10.1038/srep42591

[68] Gunduz N, Ceylan H, Guler MO, Tekinay AB. Intracellular accumulation of gold nanoparticles leads to inhibition of macropinocytosis to reduce the endoplasmic reticulum stress. Scientific Reports. 2017;**7**:40493. DOI: 10.1038/srep40493

[69] Silva E, Barreiros L, Segundo MA, Costa Lima SA, Reis S. Cellular interactions of a lipid-based nanocarrier model with human keratinocytes: Unravelling transport mechanisms. Acta Biomaterialia. 2017;**53**:439-449. DOI: 10.1016/j.actbio.2017.01.057

[70] Cao Z, Peng F, Hu Z, Chu B, Zhong Y, Su Y, et al. In vitro cellular behaviors and toxicity assays of small-sized fluorescent silicon nanoparticles. Nanoscale. 2017;**9**(22):7602-7611. DOI: 10.1039/c7nr00530j

[71] Shah P, Kaushik A, Zhu X, Zhang C, Li C-Z. Chip based single cell analysis for nanotoxicity assessment. The Analyst. 2014;**139**(9):2088-2098. DOI: 10.1039/c3an02280c

[72] Harper S, Usenko C, Hutchison JE, Maddux BLS, Tanguay RL. In vivo biodistribution and toxicity depends on nanomaterial composition, size, surface functionalisation

and route of exposure. Journal of Experimental Nanoscience. 2008;**3**(3):195-206. DOI: 10.1080/17458080802378953

[73] van Pomeren M, Brun NR, Peijnenburg WJGM, Vijver MG. Exploring uptake and biodistribution of polystyrene (nano)particles in zebrafish embryos at different developmental stages. Aquatic Toxicology. 2017;**190**:40-45. DOI: 10.1016/j.aquatox.2017.06.017

[74] Lin C-Y, Chiang C-Y, Tsai H-J. Zebrafish and Medaka: New model organisms for modern biomedical research. Journal of Biomedical Science. 2016;**23**:19. DOI: 10.1186/s12929-016-0236-5

[75] Noyes PD, Garcia GR, Tanguay RL. Zebrafish as an in vivo model for sustainable chemical design. Green Chemistry. 2016;**18**(24):6410-6430. DOI: 10.1039/C6GC02061E

[76] Bambino K, Chu J. Zebrafish in toxicology and environmental health. Current Topics in Developmental Biology. 2017;**124**:331-367. DOI: 10.1016/bs.ctdb.2016.10.007

[77] Garcia GR, Noyes PD, Tanguay RL. Advancements in zebrafish applications for 21st century toxicology. Pharmacology & Therapeutics. 2016;**161**:11-21. DOI: 10.1016/j.pharmthera.2016.03.009

[78] Zon LI, Peterson RT. In vivo drug discovery in the zebrafish. Nature Reviews. Drug Discovery. 2005;**4**(1):35-44. DOI: 10.1038/nrd1606

[79] OECD. Guidelines for the Testing of Chemicals, Section 2, Test No. 236: Fish Embryo Acute Toxicity (FET) Test. 26 July 2013:22. ISBN: 9789264203709(PDF). DOI: http://dx.doi.org/10.1787/9789264203709-en

[80] Rizzo LY, Golombek SK, Mertens ME, Pan Y, Laaf D, Broda J, et al. In vivo Nanotoxicity testing using the Zebrafish embryo assay. Journal of Materials Chemistry. B, Materials for Biology and Medicine. 2013;**1**:3918-3925. DOI: 10.1039/C3TB20528B

[81] Zhu X, Tian S, Cai Z. Toxicity assessment of iron oxide nanoparticles in Zebrafish (Danio Rerio) early life stages. PLoS One. 2012;**7**(9):e46286. DOI: 10.1371/journal.pone.0046286

[82] Park H-G, Yeo M-K. Effects of TiO_2 nanoparticles and nanotubes on zebrafish caudal fin regeneration. Molecular & Cellular Toxicology. 2013;**9**(4):375-383. DOI: 10.1007/s13273-013-0046-8

[83] de Oliveira GMT, Kist LW, Pereira TCB, Bortolotto JW, Paquete FL, de Oliveira EMN, et al. Transient modulation of acetylcholinesterase activity caused by exposure to dextran-coated iron oxide nanoparticles in brain of adult zebrafish. Comparative Biochemistry and Physiology Part C: Toxicology & Pharmacology. 2014;**162**:77-84. DOI: 10.1016/j.cbpc.2014.03.010

[84] Wang Q, Huang J-Y, Li H-Q, Chen Z, Zhao AZ-J, Wang Y, et al. TiO_2 nanotube platforms for smart drug delivery: A review. International Journal of Nanomedicine. 2016;**11**:4819-4834. DOI: 10.2147/IJN.S108847

Interaction of Nanoparticles with Blood Components and Associated Pathophysiological Effects

Gerardo González De La Cruz,
Patricia Rodríguez-Fragoso, Jorge Reyes-Esparza,
Anahí Rodríguez-López, Rocío Gómez-Cansino and
Lourdes Rodriguez-Fragoso

Additional information is available at the end of the chapter

http://dx.doi.org/10.5772/intechopen.69386

Abstract

Nanotechnology currently plays a pivotal role in several fields and has enabled substantial advances in a relatively short time. In biomedicine, nanomaterials can be potentially employed as a tool for early diagnosis and an innovative mode of drug delivery. Novel nanomaterials are currently widely manipulated without a full assessment of their potential health risks. It is commonly thought that nanomaterials' first contact with the organism is through the different components of the immune system. However, if the entry route is intravenous, the first contact will be with the blood's components (erythrocytes, platelets, white cells, plasma and complement proteins). The presence of nanomaterials within a dynamic environment such as the bloodstream can produce potential harmful effects following interaction with several blood components. The design of innovative strategies leading to the development of more hemocompatible nanomaterials is also necessary.

Keywords: nanotechnology, blood, complement, protein corona

1. Introduction

Current nanotechnology plays a pivotal role in a variety of fields and has enabled substantial advances in a relatively short time. In biomedicine, nanomaterials can be potentially employed as a tool for early diagnosis and an innovative mode of drug delivery [1–3]. As nanomaterial research grows, increased occupational exposure and very likely environmental pollution occur due to a lack of handling regulations.

The studies about the nanomaterials started fifteen years ago, and knowledge regarding their toxic potential is still limited and without appropriate regulatory measures in place [4–6]. Toxicologists, epidemiologists, and sociologists have particularly debated the future implications of nanotechnology as well as concerns regarding their toxicity and potential environmental impact. Nanomaterial application has expanded across a variety of fields, and the lack of attention involving their regulation is worrisome [7]. Novel nanomaterials are currently widely manipulated without a full assessment of their potential health risks, while nanomaterials that are already commercially available have no Safety Data Sheets. There is an urgent need for studies that will help identify, understand, and predict the cellular or tissue responses that nanomaterials can trigger in humans: it is important to understand how safe they are and establish respective protective measures.

The potential medical applications of nanomaterials are spreading to the fields of imaging, therapy, and nanodiagnostics [8, 9]. Regardless of the medical use they are given, nanomaterials come into contact with human tissues and cells which can trigger reactions that might include nanomaterial-blood interactions, damage, acute inflammation, and chronic inflammation.

Human beings have developed evolutionary defense mechanisms against the microorganisms and foreign particles with which they might potentially interact. When effective, these mechanisms provide immunity, that is, resistance to the invasive agents. Their failure results in illness. The human immune system modulates many important biological protective processes [10–12]. It coordinates responses involving a variety of cells and molecules to protect us from invading pathogens, as well as cancer cells and foreign agents [13, 14]. It is commonly thought that nanomaterials' first contact with the organism is via the different components of the immune system. However, if the entry route is intravenous, the first contact will be with the blood's components (erythrocytes, platelets, white cells, plasma and complement proteins). This interaction can lead to different associated pathophysiological processes, as will be enunciated below (**Figure 1**). Knowing what happens when nanomaterials interact with blood components is an essential step when evaluating health risks.

Until now, all biomaterials meant for use in humans trigger a tissue response when they come in contact with either healthy or sick tissue [15–17]. However, this response is triggered by physical contact between the tissue and the biomaterial which is implanted in an organ or tissue. When we talk about nanomaterials, it should be considered that due to their size, effects will take place at a nanolevel, that is, at the cellular or molecular levels. The response they trigger will not be necessarily the same as the one occurring after the implantation of a biomaterial. Additionally, since the projected applications of nanomaterials in nanomedicine involve diagnosis (imaging) and treatment (nanotransporters)—mostly for cancer—host contact with nanomaterials will be of an intravascular nature. Nanomaterial-blood interactions have been linked to inflammatory responses; early response to this damage mainly involves the blood and the vascular endothelium. Once nanomaterials enter the bloodstream, they come into contact with blood cells (red cells, white cells, and platelets), complement proteins, and plasma proteins. It is important to understand how they interact with those elements to assess their effective toxic potential, in both the blood and remote sites.

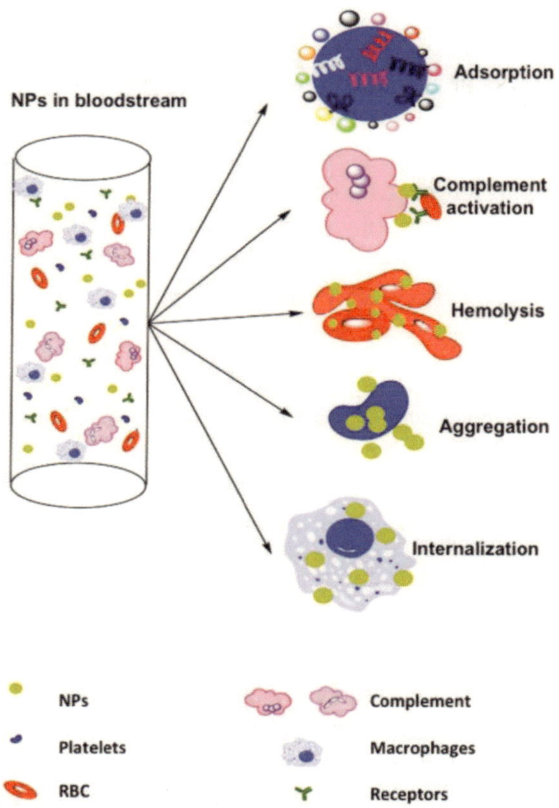

Figure 1. Interaction of nanoparticles (NPs) with bloodstream components and physiological effects.

1.1. Interaction of nanomaterials with red cells

Erythrocytes or red blood cells are exposed to attacks throughout their life span which results in constant biochemical and morphological changes. These cells' contact with nonbiological objects may significantly affect their functions [18, 19]. Nanomaterial interaction with these cells has different effects depending on their intrinsic characteristics. Venkatesan et al. [20] and Choimet et al. [21] recently reported the high hemocompatibility of chitosan nanoparticles loaded with siRNA-Npr3 and nanoparticles formed with colloidal apatite; however, chitosan nanoparticles dissolved with tripolyphosphate (TPP) to acid pH produce hemolysis [22]. Kim et al. [23] carried out rheological measurements and showed that, at concentrations of 12.5 μg/mL, silica nanomaterials caused hemolysis, deformation, and aggregation of erythrocytes. It is well known that negatively charged silver nanoagents (AgNPs) strongly interact with organic cations of the erythrocyte membrane and stimulate hemolysis [24]. Wang et al. [25] found that red blood cells and hemoglobin concentrations increased in rats treated with graphene quantum dots (5 mg/kg). The reason why erythrocytes do not show immediate damage when exposed to toxic substances is because they have a system of antioxidant defense that includes nonenzymatic antioxidants such as glutathione and antioxidant enzymes such as catalase and peroxiredoxin-2 [26, 27]. The presence of this defense system could explain their resistance to the damage induced by nanoparticles, indicating that these cells are not as sensitive to the toxic effects of nanoparticles. However, one of the factors that can influence whether a nanoparticle leads to hemolysis or not is the presence of surfactants [28] and [29] coatings. Surfactants confer different properties to nanomaterials, altering charge, interfacial tension, and becoming an amphiphilic molecule, which reduces the nanomaterial coalescence. Coatings add different chemical groups to the surface of nanoparticles confer different chemical behaviors. However, hemocompatibility will be different for each type of nanoparticle, making it impossible to anticipate if one nanoparticle will be toxic when surfactants or coatings are added.

Recent studies have shown the usefulness of erythrocytes as nanoparticle carriers: they assist their adherence to the vascular endothelium [30] and serve as a platform to bypass the immune system [31, 32]. The physical and chemical properties of erythrocytes are ideal for drug delivery. According to data regarding cell-based therapies, those using erythrocytes have proved to be the most stable, versatile, safe, and easy to manufacture. This strategy is based on temporarily opening pores in the membrane of erythrocytes, easily transporting drugs and ensuring that the latter can stay within these cells once the pores have closed [33]. One of the main uses for this system is the delivery of contrast agents contained within superparamagnetic iron oxide (SPIO) nanoparticles, ultrasmall superparamagnetic iron oxide (USPIO) nanoparticles, very small superparamagnetic iron oxide (VSPIO) nanoparticles, and monocrystalline iron oxide nanocompound (MION) particles, which are already registered and approved for use in the United States and Europe [34]. These have been successfully employed for magnetic particle inspection (MPI) techniques involving the imaging of vessels or structures filled with blood, both during interventions and when monitoring long-term cardiovascular diseases [35]. There is no doubt regarding the hemolytic potential of nanomaterials, and since the presence of hemolysis for long periods can have fatal health effects, an assessment of each nanomaterial's hemolytic potential is quite important.

1.2. Interaction of nanomaterials with platelets

As they go through the blood, nanoparticles can also interact with platelets. When these blood cells come into contact with any material, they adhere to its surface and begin a cascade of signals that eventually leads to fibrin cross-linking and clot formation. While platelet binding is the main process that maintains hemostasis and prevents hemorrhages, it can also lead to the appearance of potentially deadly thrombi that can cause strokes or cerebrovascular accidents [36]. Thrombus formation involves a series of cellular events that entail the participation of molecular components, mainly proteins. If a nanomaterial has the capacity to induce platelet aggregation and alter the normal process of coagulation, it could lead to bleeding or thrombosis. Hence, it is important to know the thrombogenic capacity of each type of nanomaterial, especially when considering their potential nanobiomedical applications.

There are few reports regarding the interaction between nanoparticles and platelets. Unfortunately, most of them show evidence of the procoagulant effects of employed nanomaterials, along with their potential health risks if the exposed person is going through pathological processes such as cardiovascular disease or metabolic syndrome. Nanoparticle-platelet interaction and endothelial injury may result in the activation of the coagulation cascade, the formation of blood clots, and the partial or total occlusion of blood vessels by thrombi. These effects will depend not only on the size, charge, hydrophobicity, or type of cover but also on the intrinsic characteristics of the nanoparticles (**Figure 2**).

Nanoparticle size seems to be an important factor for platelet activation. It has been observed that the smaller the nanoparticles are, the more likely it is that these cells will be activated. It was already demonstrated that platelet response increases with the size reduction of gold nanoparticles (AuNPs) [37]. On the other hand, SiO_2 nanoparticles have been shown to produce more severe effects when they are about 50–70 nm in size. Concordantly, another work suggests that the effect of 200 nm SiO_2 particles was more intense than that caused by 50 nm ones [38]. Nanoparticles' capacity to induce platelet activation has been amply reported. Examples include gold (AuNP) and silver (AgNP) nanoparticles, as well as CdTe and CdSe quantum dots [39–41]. As far as coating is concerned, carboxyl, amino, phosphate, or hydroxyls are all known to lead to platelet activation [42]. On the other hand, negatively charged surfaces can more easily initiate thrombotic events because, in physiological coagulation, platelet contact with anionic surface starts the coagulation cascade [43]. It was recently reported that, at low concentrations (50 ng/mL), TiO_2 nanoparticles could trigger the activation of the contact system, leading to inflammation-induced thrombosis in a complete blood in vitro model [44].

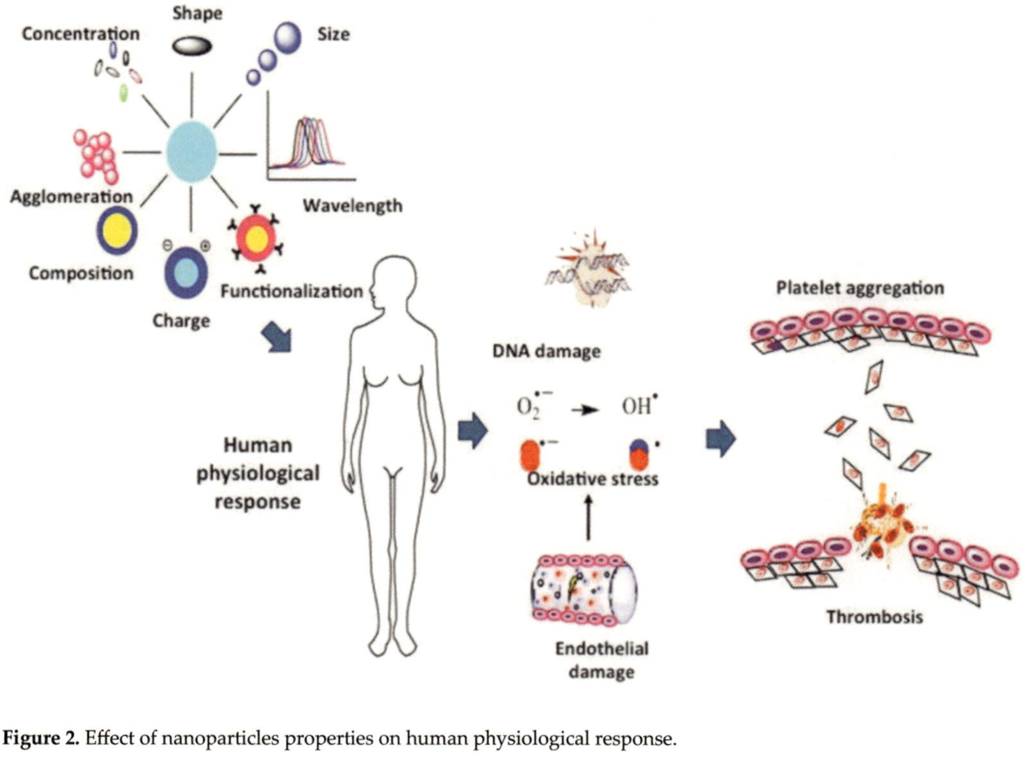

Figure 2. Effect of nanoparticles properties on human physiological response.

Simple and multiwalled carbon nanotubes (SWCNTs and MWCNTs), iron nanoparticles (Fe_2O_3 and Fe_3O_4NPs), silicon oxide nanoparticles (SiO_2 NPs), pegylated nanoparticles (PEGylated), titanium oxide nanoparticles (TiO_2NPs), and zinc oxide nanoparticles (ZnONP) are known not to induce platelet activation [45]. This makes them ideal for theranostics. Interestingly, just like several types of nanoparticles that can induce platelet activation, certain nanomaterials are being used for diagnosis in thrombotic disease since thrombogenic proteins range from nanometers to micrometers in size. High-resolution imaging is required to observe these proteins in real time, and this is being achieved via the use of nanotransporters. For example, liposomes contain thrombin inhibitors for acute thrombosis, perfluorocarbon nanoparticles with thrombin inhibitors, polymer nanotransporters with antithrombotic activity, magnetic nanoparticles conjugated with thrombolytic activity urokinase, and Fe_2O_3 nanoparticles conjugated with plasminogen activator for thrombosis [46]. Identifying the prothrombotic or nanotransporting potential of a nanomaterial is important given its potential medical implications.

1.3. Interaction of nanomaterials with peripheral mononuclear cells

Peripheral blood mononuclear cells (PBMC) are another important element in the bloodstream. While there are many studies on lymphocyte cell lines, the use of PBMC cells allows for the simultaneous analysis of nanoparticle effects on several important immune cells such as B-cells, T-cells, monocytes, and natural killer (NK) cells. Studying nanomaterials with these types of cells are crucial since these will interact with nanoparticles once the latter are introduced into the blood torrent. The peripheral blood mononuclear cells (lymphocytes and monocytes) represent a host defense system that is capable of releasing various inflammatory mediators after its activation. The physicochemical properties of nanomaterials can act as intrinsic signals that aid immunity.

Immune cells need to communicate to exercise their function. One way is via the production and release of exosomes, nanovesicles that are naturally present in and are released by the majority of cells in the body. These exosomes can act as a means of communication for transferring functional proteins, mRNA and microRNA. Studies conducted by Andersson-Willman et al. [47] show that in subtoxic concentrations, TiO_2 and ZnO nanoparticles do not interfere with the traffic and release of these exosomes in PBMC cell populations. Adhesion molecules such as integrins and selectins also play an important role in immune response. ZnO NPs alter the expression of several integrins, L-selectin and chemokine receptor CXCR4, important molecules for key cellular functions such as adhesion, migration, and cell proliferation [48]. Immune cells also have other functions, such as inducing the proliferation of other cells, phagocytizing, or killing. It has been pointed out that peripheral blood immune cells (lymphocytes, NK cells, granulocytes, and monocytes) have a different sensitivity to the effect of polylactic-glycolic acid (PLGA-PEO) nanoparticles. These particles suppressed the proliferative function of lymphocytes and the killing activity of NK cells, but stimulated phagocytic activity of granulocytes and monocytes, as well as the respiratory burst of phagocytes [49].

Nanoparticles have different toxic effects in human PBMC. Lankoff et al. [50] used silica nanoparticles and observed no significant cytotoxic and genotoxic effects in peripheral blood lymphocytes. However, cell proliferation was affected in a concentration-dependent manner. Those same authors found that surface charge and zeta potential are important for the binding and uptake of nanoparticles into cells. Peripheral blood lymphocytes died after exposure to nanoparticles containing silicon and carbon nanotubes, an effect that was dependent on the concentration of the nanoparticles employed [51, 52]. TiO_2, ZrO_2, and Al_2O_3 nanoparticles did not damage these cells' DNA at concentrations of 1, 10, and 100 μg/mL [53]. Nanoparticle size seems to influence toxicity. Studies involving peripheral blood lymphocytes treated with SiO_2 nanoparticles of different sizes (6, 20, and 50 nm) induced size-dependent cytotoxic, genotoxic, and mutagenic effects [54].

Nanoparticle coating also seems to influence the toxicity induced in peripheral blood mononuclear cells. Oleate F_3O_4 nanoparticles produce cytotoxic and genotoxic effects. Oleate seems to confer a different load and agglomeration potential, favoring cell uptake and, therefore, cytotoxicity [55]. Recently, Farace et al. studied the effect of nanocapsules (PCs, 170–300 nm) coated with chitosan and pluronic PEG on peripheral blood mononuclear cells and subpopulations of T lymphocytes and monocytes which were taken as representative of the innate immune response [56]. They observed that different types of NCs produce different effects on immune cells. For example, the PEG NCs were completely inert, while Pluro NCs and Chito NCs had immunomodulatory effects. Pluro nanoparticles induced an immune response through CD69 up-modulation in monocytes and increased the release of IL-6, IL-10, IL-12, and TNFα. On the other hand, Chito nanoparticles produced apoptosis in monocytes and T-cells, as an increase in the secretion of pro-inflammatory cytokines (TNFα and IL-12) [57]. Interestingly, Chito NCs induced the secretion of IL-4 and IL-13 cytokines. Normally, the T effector (Th_2) helper cells produce a cytokine profile that includes IL-4, IL-6, IL-10, and IL-13. These cytokines signal B-cells to proliferate and differentiate in antibody-producing plasma cells. The possible activation of Th_2 cells mediated by Chito NCs might be responsible for the induction of allergy in humans.

1.4. Interaction of nanomaterials with complement proteins

The complement system is part of innate immunity and is one of the oldest defense systems. Any absence or abnormalities in this system or any of its components can cause serious and even deadly disease. The complement system has three known modes of activation: the classical, alternative, and lectin pathways, which differ both in their activation mechanisms and initial components. The main biological functions of this system include (a) opsonization, (b) chemotaxis, (c) cellular and bacterial lysis, (d) anaphylatoxin function, and (e) participation in the elimination of immune complexes [58]. The system is composed of several components (C1, C2 to C9) and factors (B, D, HI, and P) and gets its name from the fact that it complements the immune response mediated by antibodies [59].

It is not known why certain nanoparticles cause the activation of the complement system, but it is of knowledge that the surface charge plays an important role. It has been noted that charged nanoparticles do more to activate the complement than neutral counterparts (e.g., polypropylene sulfur nanoparticles, lipid nanocapsules, polycation-based nanoparticles containing cyclodextrin and polystyrene nanospheres [60–63]). Also, polymer coatings made of polyethylene glycol (PEG) and poloxamine-908 partially neutralize the surface charge and reduce the activation of the complement system [64, 65]. However, similar studies using dextran and chitosan showed that the size and shape of the polymer influenced the degree of complement activation, which was not determined by the effects of the load [66]. The presence of thiol groups in the nanoparticle's surface also increases the activation of the complement [67].

We know that hydrophobic surfaces are more powerful activators than hydrophilic ones and that the inclusion of NH_2, –OH, or –COOH groups influences the activation of the complement [68]. Nanotubes functionalized with psychosine activate the complement via the classical pathway [69]. Several studies have shown that the activation of the complement can be dependent on size, polydispersity index, zeta potential, and surface ligand density [70–72].

The activation of the complement system leads to an inflammatory response via the release of anaphylatoxins (e.g., C3a and C5a), C3b and C5b–C9 lytic complex, since this response originates uncontrolled activation, which can lead to organism collapse [73]. One of the diseases that has been directly associated to the activation of the complement system is C activation-related pseudoallergy (CARPA), which entails reactions of hypersensitivity. Such situations demand security evaluations and the development of technologies that consider complement activation and nanomaterials' potential to induce CARPA [74]. Complement activation has also been associated with the development and growth of tumors [75]. The generation of C5a in a tumor environment increases tumor growth by promoting the recruitment of suppressor cells derived from myeloid in malignant tumors and the deregulation or suppression of CD8 cytotoxic cells. This is therefore another important consideration when evaluating new nanomaterials.

It is clear that nanomaterials can have different effects on the components of the immune system in peripheral blood and at the tissue level. It is also evident that the physical, chemical, and optical properties of nanomaterials are critical for this interaction. That is in fact what has led to the design of tools that can be employed in the field of immunology. Nanomaterials are being used in the development of vaccines and immunotherapy and even therapeutic methods that seek to inhibit the complement system in cancer treatment [76]. There is an urgent need to understand the exact mechanisms of nanoparticle/immune interaction so as to have a more complete picture of what could actually happen in humans, thus ensuring more effective and secure applications.

1.5. Interaction of nanomaterials with plasma proteins

So far, there is no evidence that just by entering the bloodstream, nanoparticles can interact with associated cells and molecules and trigger a variety of potentially adverse effects. As nanoparticles go through the systemic circulation, they will also interact with plasma proteins. Protein adsorption in biological medium is an important issue when measuring biological response to nanoparticles [77–79]. Binding of plasma components and the formation of nanoparticle-protein interactions largely determines nanoparticle fate in the systemic circulation and should influence functionality as much as the plasma protein binding of drugs [80].

Knowledge regarding nanoparticle-protein interaction mechanisms has evolved over the past few years. Initially, several authors had indicated that protein adsorption decreased subsequent to the functionalization of the nanomaterial since the hydrophobicity of the nanoparticle's surface was also reduced [81]. However, nanoparticle-protein interaction is very complex and that the scenario will depend on whether this interaction occurs in the blood, interstitial liquid, or some other biological liquid, as well as the specific surface properties of particular nanoparticles (size, shape, load, composition, and surface functionalization) [82].

It has been recently noted that once a nanoparticle enters the biological environment, it becomes coated in proteins, the so-called protein corona that will influence the fate of the nanoparticle inside the organism (i.e., the time spent in the bloodstream, biodistribution, cellular uptake, and intracellular localization) (**Figure 3**). This will also depend on the biological environment and on whether the biological environment is physiologically ill or healthy [83]. In addition to the above, it must be considered that the ultimate goal in nanomedicine is to

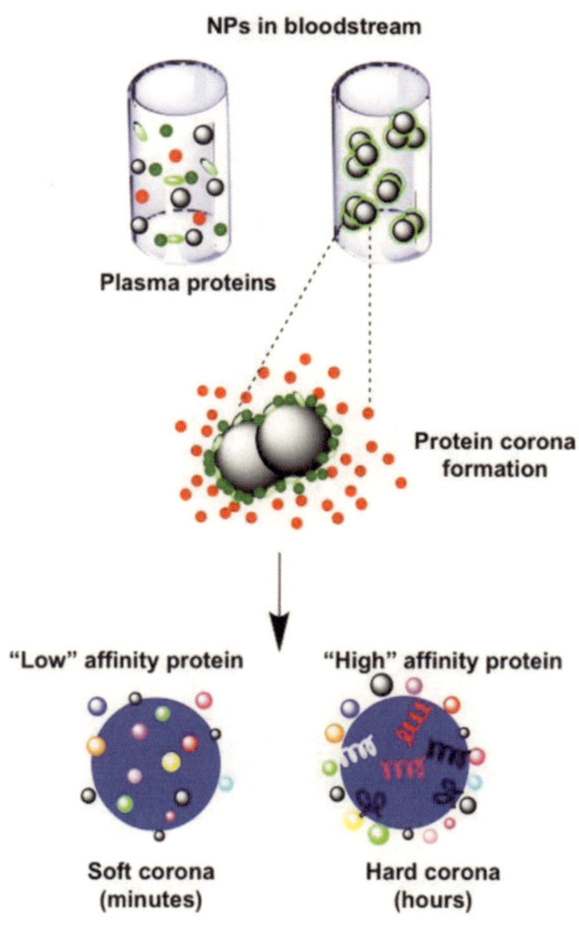

Figure 3. Nanoparticles (NPs) in bloodstream and corona protein formation.

use nanoparticles as transporters or a contrast medium for imaging. For theranostic purposes, nanoparticles must be functionalized with peptides, proteins, antibodies, oligonucleotides, or drugs, which means that the proteins that make up the corona will probably vary depending on the molecule with which the nanoparticle is conjugated and will provide the nanomaterial not only with a new biological identity but also with new physicochemical properties, changing shape, size, load, surface composition, and state of aggregation. This will also allow conformational changes [84, 85].

The binding of nanoparticles to plasma proteins such as albumin can increase biological properties that reduce the activation of the complement system, increasing blood circulation time and reducing toxicity [86]. In this regard, it would seem that the adsorption of circulating proteins might confer safety, facilitate interactions mediated by receptors, and improve the pharmacokinetic profile, i.e., these are potential theranostic advantages. However, the proteins that are involved in relevant physiological and toxicological processes in the bloodstream, such as the complement and coagulation factors, have also been identified forming the protein corona. The formation of the latter may additionally reduce target cellular interactions by making the ligands inaccessible to their surfaces [87]. For these reasons, the formation of the protein corona could be disadvantageous for theranostic purposes.

Another subject of study in this regard is that of the "hard or soft corona." The term "hard corona" describes a long life and low complexity balance, while "soft corona" consists of the formation of layers of highly complex biomolecules that exchange quickly. That said, this concept is not widely used by researchers, and studies on ligand-receptor interaction with nanoparticles do not specify if the interaction is with a hard or soft corona [88].

When does the protein corona form and how quickly? Studies by Tenzer et al. [89] showed almost instant formation upon nanoparticle contact with the blood (<30 s). This group studied positively and negatively charged polystyrene nanoparticles (nPsNPs and pPsNPS, respectively) and silica nanoparticles of various sizes (Ø ≈ 35, 120, and 140 nm), loads, and surface modifications (unmodified amine and carboxylate), exposing them to human plasma for different periods of time. They quantified the formation of 166 different protein coronas every 30 s for PsNPs and silica nanoparticles (35 nm). When modified the surfaces were able to quantify the formation of 300 different protein coronas.

Some authors have pointed out that the protein corona can increase the useful drug load capacity of nanoparticles. The very small size of the nanoparticles facilitates their travel down the bloodstream, their incorporation into cells, and their interaction with the cellular compartments with different molecules, including DNA. A way in which the protein corona can improve not only nanoparticles' drug transporting capacities but also their pharmacokinetic properties is by increasing payload capability through the use of porous materials (e.g., porous silicon) that can be packaged with the drug, as well as through the volume increase of liposomes. However, most current methods, unaware of the formation of the protein corona, typically load the drugs over the nanoparticles surface via chemical conjugation or adsorption. In these cases, the maximum amount that can be charged is restricted to a single layer [90]. Nanoparticles functionalized with polyethylene glycol (PEG) have been recently developed. PEG provides colloidal stability even under conditions of physiological salinity caused by interparticle repulsion. However, even complex pegylation is unable to completely prevent the formation of the protein corona, even though the degree of protein adsorption is clearly reduced [91].

The adsorption of serum proteins and the formation of a corona are part of the immune response. Expanding the circulation time of the nanoparticles means more contact with the blood proteins, a higher probability of thrombogenicity, and the activation of the complement system [92]. Since the corona depends on the characteristics of the nanoparticle (chemical surface, size, shape, and charge), these properties could be adjusted in such a way for the proteins that make it up to mitigate the immune response. Tuning the properties of affinity toward the corona could optimize the biocompatibility of the nanoparticles and reduce their toxicity. However, this process is not easy, and few researchers designing nanomaterials take corona formation into consideration. Ideally, nanoparticle development should encompass studies that provide data regarding nanoparticle-protein interactions, as is customary during the development of a new drug. There are few reports on this subject, and limited knowledge in this field may be a reason for the lack of successful clinical treatments.

1.6. Interaction of nanomaterials with the vascular endothelium

Oftentimes, the study of nanomaterial-blood interactions focuses on blood cells and proteins. However, the vascular endothelium where these elements are contained must also be taken into consideration and plays an important role because these cells and proteins interact with it triggering severe pathophysiological processes. In addition to the multiple direct physical interactions between nanoparticles and endothelial cells, nanoparticles enter and circulate in the blood vessel. Among other important functions, the endothelium maintains the vascular tone, vascular cell growth regulation, leukocyte and platelet adhesion regulation, thrombosis and fibrinolysis regulation, and inflammation mediation. The normal endothelium may detect hemodynamics (e.g., pressure and friction forces) and hormonal changes (e.g., vasoactive substances as well as mediators that occur in blood cells and platelets). As a consequence, endothelial cells synthesize and release biologically active substances that maintain vascular homeostasis. Endothelial damage accompanied by endothelial dysfunction plays a crucial role and is associated with a prothrombotic state. Several reports associate nanomaterial exposure to endothelial damage [93–95] because blood is the main route for nanoparticle transportation during distribution [96]. Some of the most important findings regarding nanoparticles and endothelial cell interaction are described below.

There are several reports in the literature regarding nanomaterials that affect cell viability and proliferation. Among them are gallium nitride nanoparticles (GaN NPs) [97], cerium dioxide nanoparticles (CeO_2 NPs) [98], gold nanospheres [99, 100], silica NPs [101–103], CdTe [104], and silver nanoparticles [105] to quote some examples. Silver nanoparticles have received much attention as of late due to the biological effects they produce in endothelial cells, e.g., decreased cell viability, induced apoptosis, increased ROS production, increased production of IL-6 and IL-8 interleukins, and increased expression of adhesion proteins, which can promote inflammation [106–108]. Sun et al. have shown that the interaction of silver nanoparticles with the cell membrane of endothelial cells is the main factor behind endothelial dysfunction and may be associated with thromboembolic problems [109]. It has also been found that silicon nanoparticles induce oxidative stress, inflammation, alteration of the oxide nitric (NO) balance, and endothelial dysfunction via the activation of MAPK/Nrf2. They can also induce inflammation and cytotoxicity in endothelial cells through the activation of potassium channels [103]. Moreover, it has been suggested that nanoparticles can regulate the barrier function of tight junction (occludins, claudins, and ZO proteins) because they interact with key kinases, altering not only the oxidative status around the junction between endothelial cells but also altering the blood flow into the vessel [110]. These data indicate that nanoparticle/endothelial cell interaction can modify the function of the blood vessel, whether in the site of injection at the moment of intravenous administration, during the distribution process, or when directed toward specific targets.

Nanomaterials are being used to direct and deliver drugs toward the endothelium and improve treatments for oncological, cardiovascular, pulmonary, neurologic, and ophthalmic diseases; they also have nanodiagnostic potential [111–121]. One of the problems in cancer is the formation of new blood vessels that irrigate the tumor (angiogenesis). Targeting nanoparticles toward the tumor and allowing these to exert their harmful effects on the endothelial wall could reduce the size of the tumor due to lack of nutrients. Some of the nanomaterials used for this purpose are liposomes (Ala-Pro-Arg-Pro-Gly (APRPG) liposomes). These can successfully target tumor microvessels via functionalization with Ala-Pro-Arg-Pro-Gly (APRPG) peptide, which binds the blood vessels through the VEGF receptor-1 [122]. Another application involves the treatment of corneal endothelial dystrophy to improve and integrate cell therapy through superparamagnetic nanoparticles that facilitate the delivery of human corneal endothelial cells (HCECs) [123].

2. Conclusion

To summarize, the presence of nanomaterials within a dynamic environment such as the bloodstream can produce potentially harmful effects following interaction with several blood components. As reported in the literature, results have not been wholly encouraging because the said interaction can lead to different associated physiopathological processes linked to hemocompatibility. There is no doubt that nanomaterials might have theranostic potential for different clinical specialties and that their features improve upon traditional strategies: they are small and have physicochemical and optical properties that help direct molecules toward specific sites to control specific processes on a vascular level. However, the properties behind these advantages also create limitations, since most nanomaterials can cause important nano-level interactions. The hemocompatibility of nanomaterials is essential when we consider that, regardless of the route of entry, the blood will transport them at any given time. Further in-depth studies are needed to understand, predict, and counteract the conduct of nanomaterials within the cellular and molecular microenvironments. The design of innovative strategies leading to the development of more hemocompatible nanomaterials is also necessary.

Author details

Gerardo González De La Cruz[1], Patricia Rodríguez-Fragoso[1], Jorge Reyes-Esparza[2], Anahí Rodríguez-López[2], Rocío Gómez-Cansino[2] and Lourdes Rodriguez-Fragoso[2]*

*Address all correspondence to: mrodriguezf@uaem.mx

1 Departamento de Fisica, CINVESTAV – I.P.N, Mexico, D.F., Mexico

2 Facultad de Farmacia, Universidad Autónoma del Estado de Morelos, Cuernavaca, Mexico

References

[1] Singh PK, Jairath G, Ahlawat SS. Nanotechnology: A future tool to improve quality and safety in meat industry. Journal of Food Science and Technology. 2016;**53**:1739-1749. DOI: 10.1007/s13197-015-2090-y

[2] Kour H, Malik AA, Ahmad N, Wani TA, Kaul RK, Bhat A. Nanotechnology—New lifeline for food industry. Critical Reviews in Food Science and Nutrition. 2015;**5**. DOI: 10.1080/10408398.2013.802662

[3] Bobo D, Robinson KJ, Islam J, Thurecht KJ, Corrie SR. Nanoparticle-based medicines: A review of FDA-approved materials and clinical trials to date. Pharmaceutical Research. 2016;**33**:2373-2387. DOI: 10.1007/s11095-016-1958-5

[4] Fard JK, Jafari S, Eghbal MA. A review of molecular mechanisms involved in toxicity of nanoparticles. Advanced Pharmaceutical Bulletin. 2015;**5**:447-454. DOI: 10.15171/apb.2015.061

[5] Haase A, Tentschert J, Luch A. Nanomaterials: A challenge for toxicological risk assessment?. EXS. 2012;**101**:219-250. DOI: 10.1007/978-3-7643-8340-4-8

[6] Hulla JE, Sahu SC, Hayes AW. Nanotechnology: History and future. Human & Experimental Toxicology. 2015;**34**:1318-1321. DOI: 10.1177/0960327115603588

[7] Mühlebach S, Borchard G, Yildiz S. Regulatory challenges and approaches to characterize nanomedicines and their follow-on similars. Nanomedicine (London, England). 2015;**10**:659-674. DOI: 10.2217/nnm.14.189

[8] Keles E, Song Y, Du D, Dong WJ, Lin Y. Recent progress in nanomaterials for gene delivery applications. Biomaterials Science. 2016;**4**:1291-1309. DOI: 10.1039/c6bm00441e

[9] Padmanabhan P, Kumar A, Kumar S, Chaudhary RK, Gulyás B. Nanoparticles in practice for molecular-imaging applications: An overview. Acta Biomaterialia. 2016;**41**:1-16. DOI: 10.1016/j.actbio.2016.06.003

[10] Walls J, Sinclair L, Finlay D. Nutrient sensing, signal transduction and immune responses. Seminars in Immunology. 2016;**28**:396-407. DOI: 10.1016/j.smim.2016.09.001

[11] Kulkarni OP, Lichtnekert J, Anders HJ, Mulay SR. The immune system in tissue environments regaining homeostasis after injury: Is "Inflammation" always inflammation?. Mediators of Inflammation. 2016;**2016**:2856213. DOI: 10.1155/2016/2856213

[12] Weidenbusch M, Anders HJ. Tissue microenvironments define and get reinforced by macrophage phenotypes in homeostasis or during inflammation, repair and fibrosis. Journal of Innate Immunity. 2012;**4**:463-477. DOI: 10.1159/000336717

[13] Plitas G, Rudensky AY. Regulatory T cells: Differentiation and function. Cancer Immunology Research. 2016 Sep 2;**4**:721-725. DOI: 10.1158/2326-6066

[14] Guillerey C, Huntington ND, Smyth MJ. Targeting natural killer cells in cancer immunotherapy. Nature Immunology. 2016;**17**:1025-1036. DOI: 10.1038/ni.3518

[15] Anderson JM. Biological responses to materials. Annual Review of Materials Research. 2001;**31**:81-110. DOI: 10.1146/annurev.matsci.31.1.81

[16] Gardner AB, Lee SKC, Woods EC, and Acharya AP. Biomaterials-based modulation of the immune system. BioMed Research International. 2013;**2013**:732182. DOI: 10.1155/2013/732182

[17] Christo SN, Diener KR, Bachhuka A, Vasilev K, Hayball JD. Innate immunity and biomaterials at the nexus: Friends or foes. BioMed Research International. 2015;**2015**:342304. DOI: 10.1155/2015/342304

[18] Mendonça R, Silveira AA, Conran N. Red cell DAMPs and inflammation. Inflammation Research. 2016;**65**:665-678. DOI: 10.1007/s00011-016-0955-9

[19] Manaargadoo-Catin M, Ali-Cherif A, Pougnas JL, Perrin C. Hemolysis by surfactants–A review. Advance in Colloid and Interface Science. 2016;**228**:1-16. DOI: 10.1016/j.cis.2015.10.011

[20] Venkatesan B, Tumala A, Subramanian V, Vellaichamy E. Data on synthesis and characterization of chitosan nanoparticles for in vivo delivery of siRNA-Npr3: Targeting NPR-C expression in the heart. Data in Brief. 2016;**8**:441-447. DOI: 10.1016/j.dib.2016.05.074

[21] Choimet M, Hyoung-Mi K, Jae-Min O, Tourrette A, Drouet C. Nanomedicine: Interaction of biomimetic apatite colloidal nanoparticles with human blood components. Colloids and Surfaces B: Biointerfaces. 2016;**145**:87-94. DOI: 10.1016/j.colsurfb.2016.04.038

[22] de Lima JM, Sarmento RR, de Souza JR, Brayner FA, Feitosa AP, Padilha R, Alves LC, Porto IJ, Batista RF, de Oliveira JE, de Medeiros ES, Bonan PR, Castellano LR. Evaluation of hemagglutination activity of chitosan nanoparticles using human erythrocytes. BioMed Research International. 2015;**2015**:247965. DOI: 10.1155/2015/247965

[23] Kim J, Heo YJ, Shin S. Haemocompatibility evaluation of silica nanomaterials using hemorheological measurements. Clinical Hemorheology and Microcirculation. 2016;**62**:99-107. DOI: 10.3233/CH-151953

[24] Chen LQ, Fang L, Ling J, Ding CZ, Kang B, Huang CZ. Nanotoxicity of silver nanoparticles to red blood cells: Size dependent adsorption, uptake, and hemolytic activity. Chemical Research in Toxicology. 2015;**28**:501-509. DOI: 10.1021/tx500479m

[25] Wang K, Gao Z, Gao G, Wo Y, Wang Y, Shen G, Cui D. Systematic safety evaluation on photoluminiscent carbon dots. Nanoscale Research Letters. 2013;**8**:122. DOI: 10.1186/1556-276X-8-122

[26] Lee T, Kim S, Yu S, Kim S, Park D, Moon H. Peroxiredoxin II is essential for sustaining life span of erythrocytes in mice. Blood. 2003;**101**:5033-5038. DOI: 10.1182/blood-2002-08-2548

[27] Gonzalez R, Auclair C, Voisin E, Gautero H, Dhermy D, and Boivin P. Superoxide dismutase, catalase, and glutathione peroxidase in red blood cells from patients with malignant disease. Cancer Research. 1984;**44**:4137-4139

[28] Thasneem YM, Sajeesh S, Sharma CP. Effect of thiol functionalization on the hemocompatibility of PLGA nanoparticles. Journal of Biomedical Materials Research. Part A. 2011;**99**:607-617. DOI: 10.1002/jbm.a.33220

[29] Barshtein G, Livshits L, Shvartsman LD, Shlomai NO, Yedgar S, Arbell D. Polystyrene nanoparticles activate erythrocyte aggregation and adhesion to endothelial cells. Cell Biochemistry and Biophysics. 2016;**74**:19-27. DOI: 10.1007/s12013-015-0705-6

[30] Villa CH, Pan DC, Zaitsev S, Cines DB, Siegel DL, Muzykantov VR. Delivery of drugs bound to erythrocytes: New avenues for an old intravascular carrier. Therapeutic Delivery. 2015;**6**:795-826. DOI: 10.4155/tde.15.34

[31] Villa CH, Anselmo AC, Mitragotri S, Muzykantov V. Red blood cells: Supercarriers for drugs, biologicals, and nanoparticles and inspiration for advanced delivery systems. Advanced Drug Delivery Reviews. 2016;**106**:88-103. DOI: 10.1016/j.addr.2016.02.007

[32] Zhang H. Erythrocytes in nanomedicine: An optimal blend of natural and synthetic materials. Biomaterials Science. 2016;**4**:1024-1031. DOI: 10.1039/c6bm00072j

[33] Rossi L, Pierigè F, Antonelli A, Bigini N, Gabucci C, Peiretti E, Magnani M. Engineering erythrocytes for the modulation of drugs' and contrasting agents' pharmacokinetics and biodistribution. Advanced Drug Delivery Reviews. 2016;**106**:73-87. DOI: 10.1016/j.addr.2016.05.008

[34] Takeuchi Y, Suzuki H, Sasahara H, Ueda J, Yabata I, Itagaki K, Saito S, Murase K. Encapsulation of iron oxide nanoparticles into red blood cells as a potential contrast agent for magnetic particle imaging. Advanced Biomedical Engineering. 2014;**3**:3743. DOI: 10.14326/abe.3.37

[35] Ekstrand-Hammarström B, Hong J, Davoodpour P, Sandholm K, Ekdahl KN, Bucht A, Nilsson B. TiO_2 nanoparticles tested in a novel screening whole human blood model of toxicity trigger adverse activation of the kallikrein system at low concentrations. Biomaterials. 2015;**51**:58-68. DOI: 10.1016/j.biomaterials.2015.01.031

[36] Laloy J, Minet V, Alpan L, Mullier F, Beken S, Toussaint O, Lucas S, Dogné J. Impact of silver nanoparticles on haemolysis, platelet function and coagulation. Nanobiomedicine. 2014;**4**:1-9. DOI: 10.5772/59346

[37] Deb S, Patra HK, Lahiri P, Dasgupta AK, Chakrabarti K, Chaudhuri U. Multistability in platelets and their response to gold nanoparticles. Nanomedicine. 2011;**7**:376-384

[38] Nemmar A, Albarwani S, Beegam S, Yuvaraju P, Yasin J, Attoub S, Ali BH. Amorphous silica nanoparticles impair vascular homeostasis and induce systemic inflammation. International Journal of Nanomedicine. 2014;**9**:2779-2789. DOI: 10.2147/IJN.S52818

[39] Aseychev AV, Azizova OA, Beckman EM, Dudnik LB, Sergienko VI. Effect of gold nanoparticles coated with plasma components on ADP-induced platelet aggregation. Bulletin of Experimental Biology and Medicine. 2013;**155**:685-688

[40] Samuel SP, Santos-Martinez MJ, Medina C, Jain N, Radomski MW, Prina-Mello A, Volkov Y. CdTe quantum dots induce activation of human platelets: Implications for nanoparticle hemocompatibility. International Journal of Nanomedicine. 2015;**10**:2723-2734. DOI: 10.2147/IJN.S78281

[41] Dunpall R, Nejo AA, Pullabhotla VS, Opoku AR, Revaprasadu N, Shonhai A. An in vitro assessment of the interaction of cadmium selenide quantum dots with DNA, iron, and blood platelets. IUBMB Life. 2012;**64**:995-1002. DOI: 10.1002/iub.1100

[42] Fröhlich E. Action of Nanoparticles on platelet activation and plasmatic coagulation. Current Medicinal Chemistry. 2016;**23**:408-430

[43] Ilinskaya AN, Dobrovolskaia MA. Nanoparticles and the blood coagulation system. Part I: Benefits of nanotechnology. Nanomedicine (London, England). 2013;**8**:773-784. DOI: 10.2217/nnm.13.48

[44] Nemmar A, Melghit K, AlI BH. The acute proinflammatory and prothrombotic effects of pulmonary exposure to rutile TiO_2 nanorods in rats. Experimental Biology and Medicine (Maywood, NJ). 2008;**233**:610-619. DOI: 10.3181/0706-RM-165

[45] Karagkiozaki V, Pappa F, Arvaniti D, Moumkas A, Konstantinou D, Logothetidis S. The melding of nanomedicine in thrombosis imaging and treatment: A review. Future Science OA. 2016;**2**:FSO113. DOI: 10.4155/fso.16.3

[46] Varna M, Juenet M, Bayles R, Mazighi M, Chauvierre C, Letourneu D. Nanomedicine as a strategy to fight thrombotic diseases. Future Science OA. 2015;**1**:4. DOI: 10.4155/fso.15.46

[47] Andersson-Willman B, Gehrmann U, Cansu Z, Buerki-Thurnherr T, Krug HF, Gabrielsson S, Scheynius A. Effects of subtoxic concentrations of TiO_2 and ZnO nanoparticles on human lymphocytes, dendritic cells and exosome production. Toxicology and Applied Pharmacology. 2011;**264**:94-103. DOI: 10.1016/j.taap.2012.07.021

[48] Lozano-Fernández T, Ballester-Antxordoki L, Pérez-Temprano N, Rojas E, Sanz D, Iglesias-Gaspar M, Moya S, González-Fernández Á, Rey M. Potential impact of metal oxide nanoparticles on the immune system: The role of integrins, L-selectin and the chemokine receptor CXCR4. Nanomedicine. 2014;**10**:1301-1310. DOI: 10.1016/j.nano.2014.03.007

[49] Tulinska J, Kazimirova A, Kuricova M, Barancokova M, Liskova A, Neubauerova E, Drlickova M, Ciampor F, Vavra I, Bilanicova D, Pojana G, Staruchova M, Horvathova M, Jahnova E, Volkovova K, Bartusova M, Cagalinec M, Dusinska M. Immunotoxicity

and genotoxicity testing of PLGA-PEO nanoparticles in human blood cell model. Nanotoxicology. 2015;**9**:1:33-43. DOI: 10.3109/17435390.2013.816798

[50] Lankoff A, Arabski M, Wegierek-Ciuk A, Kruszewski M, Lisowska H, Banasik-Nowak A, Rozga-Wijas K, Wojewodzka M, Slomkowski S. Effect of surface modification of silica nanoparticles on toxicity and cellular uptake by human peripheral blood lymphocytes in vitro. Nanotoxicology. 2013;**7**:235-250. DOI: 10.3109/17435390.2011.649796

[51] Andreeva ER, Rudimov EG, Gornostaeva AN, Beklemyshev VI, Makhonin II, Maugeri UO, Buravkova LB. In vitro study of interactions between silicon-containing nanoparticles and human peripheral blood leukocytes. Bulletin of Experimental Biology and Medicine. 2013;**155**:396-398

[52] Kim JS, Song KS, Yu IJ. Multiwall carbon nanotube-induced DNA damage and cytotoxicity in male human peripheral blood lymphocytes. International Journal of Toxicology. 2016 Jan-Feb;**35**(1):27-37. DOI: 10.1177/1091581815598749

[53] Demir E, Burgucu D, Turna F, Aksakal S, Kaya B. Determination of TiO_2, ZrO_2, and Al_2O_3 nanoparticles on genotoxic responses in human peripheral blood lymphocytes and cultured embyronic kidney cells. Journal of Toxicology and Environmental Health A. 2013;**76**:990-1002. DOI: 10.1080/15287394.2013.830584

[54] Battal D, Çelik A, Güler G, Aktaş A, Yildirimcan S, Ocakoglu K, Çömelekoğlu Ü. SiO_2 Nanoparticule-induced size-dependent genotoxicity – An in vitro study using sister chromatid exchange, micronucleus and comet assay. Drugs and Chemical Toxicology. 2015;**38**:196-204. DOI: 10.3109/01480545.2014.928721

[55] Magdolenova Z, Drlickova M, Henjum K, Rundén-Pran E, Tulinska J, Bilanicova D, Pojana G, Kazimirova A, Barancokova M, Kuricova M, Liskova A, Staruchova M, Ciampor F, Vavra I, Lorenzo Y, Collins A, Rinna A, Fjellsbø L, Volkovova K, Marcomini A, Amiry-Moghaddam M, Dusinska M. Coating-dependent induction of cytotoxicity and genotoxicity of iron oxide nanoparticles. Nanotoxicology. 2015;**9**:1:44-56. DOI: 10.3109/17435390.2013.847505

[56] Farace C, Sánchez-Moreno P, Orecchioni M, Manetti R, Sgarrella F, Asara Y, Peula-García JM, Marchal JA, Madeddu R, Delogu LG. Immune cell impact of three differently coated lipid nanocapsules: Pluronic, chitosan and polyethylene glycol. Scientific Reports. 2016;**6**:18423. DOI: 10.1038/srep18423

[57] Khon X, Hellerman GR, Zhang W, Jena P, Kumar M, Behera A, Behera S, Lockey R, Mohapatra SS. Chitosan interferon-γ nanogene therapy for lung disease: Modulation of t-cell and dendritic cell immune responses. Allergy, Asthma & Clinical Immunology. 2008;**4**:95. DOI: 10.1186/1710-1492-4-4-95

[58] Pham CT, Mitchell LM, Huang JL, Lubniewski CM, Schall OF, Killgore JK, Pan D, Wickline SA, Lanza GM, Hourcade DE. Variable antibody-dependent activation of complement by functionalized phospholipid nanoparticle surfaces. Journal of Biological Chemistry. 2011;**286**:123-130. DOI: 10.1074/jbc.M110.180760

[59] Knopf PM, Rivera DS, Hai SH, McMurry J, Martin W, De Groot AS. Novel function of complement C3d as an autologous helper T-cell target. Immunology & Cell Biology. 2008;**86**:221-225

[60] Reddy ST, van der Vlies AJ, Simeoni E, Angeli V, Randolph GJ, O'Neil CP, Lee LK, Swartz MA, Hubbell JA. Exploiting lymphatic transport and complement activation in nanoparticle vaccines. Nature Biotechnology. 2007;**25**:1159-1164. DOI: 10.1038/nbt1332

[61] Vonarbourg A, Passirani C, Saulnier P, Simard P, Leroux JC, Benoit JP. Evaluation of pegylated lipid nanocapsules versus complement system activation and macrophage uptake. Journal of Biomedical Materials Research. Part A. 2006;**78**:620-628. DOI: 10.1002/jbm.a.30711

[62] Bartlett DW, Davis ME. Physicochemical and biological characterization of targeted, nucleic acid-containing nanoparticles. Bioconjugate Chemistry. 2007;**18**:456-468. DOI: 10.1021/bc0603539

[63] Nagayama S, Ogawara K, Fukuoka Y, Higaki K, Kimura T. Time-dependent changes in opsonin amount associated on nanoparticles alter their hepatic uptake characteristics. International Journal of Pharmaceutics. 2007;**342**:215-221

[64] Al-Hanbali O, Rutt KJ, Sarker DK, Hunter AC, Moghimi SM. Concentration dependent structural ordering of poloxamine 908 on polystyrene nanoparticles and their modulatory role on complement consumption. Journal of Nanoscience and Nanotechnology. 2006;**6**:3126-3133

[65] Bertholon I, Vauthier C, Labarre D. Complement activation by core-shell poly(isobutylcyanoacrylate)-polysaccharide nanoparticles: Influences of surface morphology, length, and type of polysaccharide. Pharmaceutical Research. 2006;**23**:1313-1323. DOI: 10.1007/s11095-006-0069-0

[66] Talaei F, Azhdarzadeh M, Hashemi Nasel H, Moosavi M, Foroumadi A, Dinarvand R, Atyabi F. Core shell methyl methacrylate chitosan nanoparticles: In vitro mucoadhesion and complement activation. Daru. 2011;**19**:257-265. DOI: 10.1038/srep02814

[67] Nilsson KN, Ekdahl TE, Mollnes JD. Lambris, The role of complement in biomaterial-induced inflammation. Molecular Immunology. 2007;**44**:82-94. DOI: 10.1016/j.molimm.2006.06.020

[68] Rybak-Smith MJ, Tripisciano C, Borowiak-Palen E, Lamprecht C, Sim RB. Effect of functionalization of carbon nanotubes with psychosine on complement activation and protein adsorption. Journal of Biomedical Nanotechnology. 2011;**7**:830-839

[69] Pacheco PM, LE B, White D, Sulchek T. Tunable complement activation by particles with variable size and Fc density. Nano Life. 2013;**3**:1341001. DOI: 10.1142/S1793984413410018

[70] Yu K, Lai BF, Foley JH, Krisinger MJ, Conway EM, Kizhakkedathu JN. Modulation of complement activation and amplification on nanoparticle surfaces by glycopolymer conformation and chemistry. ACS Nano. 2014;**8**:7687-7703. DOI: 10.1021/nn504186b

[71] Thomas DG, Chikkagoudar S, Heredia-Langer A, Tardiff MF, Xu Z, Hourcade DE, Pham CT, Lanza GM, Weinberger KQ, Baker NA. Physicochemical signatures of nanoparticle-dependent complement activation. Computational Science & Discovery. 2014;**7**:015003. DOI: 10.1088/1749-4699/7/1/015003

[72] Banda NK, Mehta G, Chao Y, Wang G, Inturi S, Fossati-Jimack L, Botto M, Wu L, Moghimi SM, Simberg D. Mechanisms of complement activation by dextran-coated superparamagnetic iron oxide (SPIO) nanoworms in mouse versus human serum. Particle and Fibre Toxicology. 2014;**11**:64. DOI: 10.1186/s12989-014-0064-2

[73] Cho H. Complement regulation: Physiology and disease relevance. Korean Journal of Pediatrics. 2015;**58**:239-244. DOI: 10.3345/kjp.2015.58.7.239

[74] Szebeni J, Storm G. Complement activation as a bioequivalence issue relevant to the development of generic liposomes and other nanoparticulate drugs. Biochemical and Biophysical Research Communications. 2015;**468**:490-497. DOI: 10.1016/j.bbrc.2015.06.177

[75] Moghimi SM. Cancer nanomedicine and the complement system activation paradigm: Anaphylaxis and tumour growth. Journal of Controlled Release. 2014;**190**:556-562. DOI: 10.1016/j.jconrel.2014.03.051

[76] Hotaling NA, Tang L, Irvine DJ, Babensee JE. Biomaterial strategies for immunomodulation. Annual Review of Biomedical Engineering. 2015;**17**:317-349. DOI: 10.1146/annurev-bioeng-071813-104814

[77] Cedervall T, Lynch I, Lindman S, Berggård T, Thulin E, Nilsson H, Dawson KA, Linse S. Understanding the nanoparticle–protein corona using methods to quantify exchange rates and affinities of proteins for nanoparticles. Proceedings of the National Academy of Sciences of the United States of America. 2007;**104**:2050-2055. DOI: 10.1073/pnas.0608582104

[78] Chen K, Rana S, Moyano DF, Xu Y, Guo X, Rotello VM. Optimizing the selective recognition of protein isoforms through tuning of nanoparticle hydrophobicity. Nanoscale. 2014;**6**:6492-6495. DOI: 10.1039/c4nr01085j

[79] Lundqvist M, Sethson I, Jonsson BH. Protein adsorption onto silica nanoparticles: Conformational changes depend on the particles' curvature and the protein stability. Langmuir. 2004;**20**:10639-10647. DOI: 10.1021/la0484725

[80] Lynch I, Dawson KA. Protein-nanoparticle interactions. Nanotoday. 2008;**3**:40-47. DOI: 10.1016/S1748-0132(08)70014-8

[81] Saptarshi SR, Duschl A, Lopata AL. Interaction of nanoparticles with proteins: Relation to bio-reactivity of the nanoparticle. Journal of Nanbiotechnology. 2013;**11**:26. DOI: 10.1186/1477-3155-11-26

[82] Treuel L, Brandholt S, Maffre P, Wiegele S, Shang L, Nienhaus GU. Impact of protein modification on the protein corona on nanoparticles and nanoparticle-cell interactions. ACS Nano. 2014;**8**:503-513. DOI: 10.1021/nn405019v

[83] Maffre P, Brandholt S, Nienhaus K, Shang L, Parak WJ, Nienhaus GU. Effects of surface functionalization on the adsorption of human serum albumin onto nanoparticles – A fluorescence correlation spectroscopy study. Beilstein Journal of Nanotechnology. 2014;**5**:2036-2047. DOI: 10.3762/bjnano.5.212

[84] Caracciolo G, Farokhzad OC, Mahmoudi M. Biological identity of nanoparticles in vivo: Clinical implications of the protein corona. Trends in Biotechnology. 2016; **S0167-7799**:30149-30154. DOI: 10.1016/j.tibtech.2016.08.011

[85] Monopoli MP, Aberg C, Salvati A, Dawson KA. Biomolecular coronas provide the biological identity of nanosized materials. Nature Nanotechnology. 2012;**7**:779-786. DOI: 10.1038/nnano.2012.207

[86] Peng Q, Zhang S, Yang Q, Zhang T, Wei XQ, Jiang L, Zhang CL, Chen QM, Zhang ZR, Lin YF. Preformed albumin corona, a protective coating for nanoparticles based drug delivery system. Biomaterials. 2013;**34**:8521-8530. DOI: 10.1016/j.biomaterials.2013.07.102

[87] Pearson RM, Juetter VV, Hong S. Biomolecular corona on nanoparticles: A survey of recent literature and its implications in targeted drug delivery. Frontiers in Chemistry. 2014;**2**:108. DOI: 10.3389/fchem.2014.00108

[88] Sacchetti C, Motamedchaboki K, Magrini A, Palmieri G, Mattei M, Bernardini S, Rosato N, Bottini N, Bottini M. Surface polyethylene glycol conformation influences the protein corona of polyethylene glycol-modified single-walled carbon nanotubes: Potential implications on biological performance. ACS Nano. 2013;**7**:1974-1989. DOI: 10.1021/nn400409h

[89] Tenzer S, Docter D, Kuharev J, Musyanovych A, Fetz V, Hecht R, Schlenk F, Fischer D, Kiouptsi K, Reinhardt C, Landfester K, Schild H, Maskos M, Knauer SK, Stauber RH. Rapid formation of plasma protein corona critically affects nanoparticle pathophysiology. Nature Nanotechnology. 2013;**8**:772-781. DOI: 10.1038/nnano.2013.181

[90] Hamad-Schifferli K. Exploiting the novel properties of protein coronas: Emerging applications in nanomedicine. Nanomedicine (London, England). 2015;**10**:1663-1674. DOI: 10.2217/nnm.15.6

[91] Docter D, Westmeier D, Markiewicz M, Stolte S, Knauer SK, Stauber RH. The nanoparticle biomolecule corona: Lessons learned - challenge accepted? Chemical Society Reviews. 2015;**44**:6094-6121. DOI: 10.1039/c5cs00217f

[92] Lee YK, Choi EJ, Webster TJ, Kim SH, Khang D. Effect of the protein corona on nanoparticles for modulating cytotoxicity and immunotoxicity. International Journal of Nanomedicine. 2014;**10**:97-113. DOI: 10.2147/IJN.S72998

[93] Montiel-Dávalos A, Ventura-Gallegos JL, Alfaro-Moreno E, Soria-Castro E, García-Latorre E, Cabañas-Moreno JG, del Pilar Ramos-Godinez M, López-Marure R. TiO_2 nanoparticles induce dysfunction and activation of human endothelial cells. Chemical Research in Toxicology. 2012;**25**:920-930. DOI: 10.1021/tx200551u

[94] Liu X, Xue Y, Ding T, Sun J. Enhancement of proinflammatory and procoagulant responses to silica particles by monocyte-endothelial cell interactions. Particle and Fibre Toxicology. 2012;**9**:36. DOI: 10.1186/1743-8977-9-36

[95] Gaillet S, Rouanet JM. Silver nanoparticles: Their potential toxic effects after oral exposure and underlying mechanisms--A review. Food and Chemical Toxicology. 2015;**77**:58-63. DOI: 10.1016/j.fct.2014.12.019

[96] Gromnicova R, Kaya M, Romero IA, Williams P, Satchell S, Sharrack B, Male D. Transport of gold nanoparticles by vascular endothelium from different human tissues. PLoS One. 2016;**11**:e0161610. DOI: 10.1371/journal.pone.0161610

[97] Braniste T, Tiginyanu I, Horvath T, Raevschi S, Cebotari S, Lux M, Haverich A, Beilstein AH. Viability and proliferation of endothelial cells upon exposure to GaN nanoparticles. Beilstein Journal of Nanotechnology. 2016;**7**:1330-1337. DOI: 10.3762/bjnano.7.124

[98] Strobel C, Oehring H, Herrmann R, Förster M, Reller A, Hilger I. Fate of cerium dioxide nanoparticles in endothelial cells: Exocytosis. Journal of Nanoparticle Research. 2015;**17**:206. DOI: 10.1007/s11051-015-3007-4

[99] Freese C, Uboldi C, Gibson MI, Unger RE, Weksler BB, Romero IA, Couraud PO, Kirkpatrick CJ. Uptake and cytotoxicity of citrate-coated gold nanospheres: Comparative studies on human endothelial and epithelial cells. Particle and Fibre Toxicology. 2012;**9**:23. DOI: 10.1186/1743-8977-9-23

[100] Anspach L, Unger RE, Gibson MI, Klok HA, Kirkpatrick CJ, Freese C. Impact of polymer-modified gold nanoparticles on brain endothelial cells: Exclusion of endoplasmic reticulum stress as a potential risk factor. Nanotoxicology. 2016;**10**:1341-1350. DOI: 10.1080/17435390.2016.1214761

[101] Freese C, Schreiner D, Anspach L, Bantz C, Maskos M, Unger RE, Kirkpatrick CJ. In vitro investigation of silica nanoparticle uptake into human endothelial cells under physiological cyclic stretch. Particle and Fibre Toxicology. 2014;**11**:68. DOI: 10.1186/s12989-014-0068-y

[102] Yang L, Yan Q, Zhao J, Li J, Zong X, Yang L, Wang Z. The role of potassium channel in silica nanoparticle-induced inflammatory effect in human vascular endothelial cells in vitro. Toxicology Letters. 2013;**223**:16-24. DOI: 10.1016/j.toxlet.2013.08.017

[103] Guo C, Xia Y, Niu P, Jiang L, Duan J, Yu Y, Zhou X, Li Y, Sun Z. Silica nanoparticles induce oxidative stress, inflammation, and endothelial dysfunction in vitro via activation of the MAPK/Nrf2 pathway and nuclear factor-κB signaling. International Journal of Nanomedicine. 2015;**10**:1463-1477. DOI: 10.2147/IJN.S76114

[104] Yan M, Zhang Y, Xu K, Fu T, Qin H, Zheng X. An in vitro study of vascular endothelial toxicity of CdTe quantum dots. Toxicology. 2011;**282**:94-103. DOI: 10.1016/j.tox.2011.01.015

[105] Rosas-Hernandez H, Jimenez-Badillo S, Martinez-Cuevas PP, Gracia- Espino E, Terrones H, Terrones M, Hussain SM, Ali SF, Gonzalez C. Effects of 45-nm silver nanoparticles on coronary endothelial cells and isolated rat aortic rings. Toxicology Letters. 2009;**191**:305-313. DOI: 10.1016/j.toxlet.2009.09.014

[106] Shi J, Sun X, Lin Y, Zou X, Li Z, Liao Y, Du M, Zhang H. Endothelial cell injury and dysfunction induced by silver nanoparticles through oxidative stress via IKK/NF-κB pathways. Biomaterials. 2014;**35**:6657-6666. DOI: 10.1016/j.biomaterials.2014.04.093

[107] Trickler WJ, Lantz SM, Murdock RC, Schrand AM, Robinson BL, Newport GD, Schlager JJ, Oldenburg SJ, Paule MG, Slikker W Jr, Hussain SM, Ali SF. Silver nanoparticle induced blood–brain barrier inflammation and increased permeability in primary rat brain microvessel endothelial cells. Toxicological Sciences. 2010;**118**:160-170. DOI: 10.1093/toxsci/kfq244

[108] Grosse S, Evje L, Syversen T. Silver nanoparticle-induced cytotoxicity in rat brain endothelial cell culture. Toxicology In Vitro. 2013;**27**:305-313

[109] Sun X, Shi J, Zou X, Wang C, Yang Y, Zhang H. Silver nanoparticles interact with the cell membrane and increase endothelial permeability by promoting VE-cadherin internalization. Journal of Hazardous Materials. 2016;**317**:570-578. DOI: 10.1016/j.jhazmat.2016.06.023

[110] Zhang E, Yang WX. Tight junction between endothelial cells: The interaction between nanoparticles and blood vessels. Beilstein Journal of Nanotechnology. 2016;**7**:675-684. DOI: 10.3762/bjnano.7.60

[111] Cheng Y, Morshed R, Auffinger B, Tobias AL, Lesniak MS. Multifunctional nanoparticles for brain tumor imaging and therapy. Advanced Drug Delivery Reviews. 2014;**66**:42-57. DOI: 10.1016/j.addr.2013.09.006

[112] Asai T. Nanoparticle-mediated delivery of anticancer agents to tumor angiogenic vessels. Biological and Pharmaceutical Bulletin. 2012;**35**:1855-1861

[113] Chen X, Fang J, Wang S, Liu H, Du X, Chen J, Li X, Yang Y, Zhang B, Zhang W. A new mosaic pattern in glioma vascularization: Exogenous endothelial progenitor cells integrating into the vessels containing tumor-derived endothelial cells. Oncotarget. 2014;**5**:1955-1968. DOI: 10.18632/oncotarget.1885

[114] Kim YT, Lobatto ME, Kawahara T, Chung BL, Mieszawska AJ, Sanchez-Gaytan BL, Fay F, Senders ML, Calcagno C, Becraft J, Saung MT, Gordon RE, Stroes ESG, Ma M, Farokhzad OC, Fayad ZA, Mulder WJM, Langer R. Probing nanoparticle translocation across the permeable endothelium in experimental atherosclerosis. Proceedings of the National Academy of Sciences of the United States of America. 2014;**111**:1078-1083. DOI: 10.1073/pnas.1322725111

[115] Agyare E, Kandimalla K. Delivery of polymeric nanoparticles to target vascular diseases. Journal of Biomolecular Research & Therapeutics. 2014;**3**:S1-S001

[116] Shuvaev VV, Christofidou-Solomidou M, Scherpereel A, Simone E, Arguiri E, Tliba S, Pick J, Kennel S, Albelda SM, Muzykantov VR. Factors modulating the delivery and effect of enzymatic cargo conjugated with antibodies targeted to the pulmonary endothelium. Journal of Controlled Release. 2007;**118**:235-244. DOI: 10.1016/j.jconrel.2006.12.025

[117] Chen C, Mei H, Shi W, Deng J, Zhang B, Guo T, Wang H, Hu Y. EGFP-EGF1-conjugated PLGA nanoparticles for targeted delivery of siRNA into injured brain microvascular endothelial cells for efficient RNA interference. PLoS One. 2013;**8**:e60860. DOI: 10.1371/journal.pone.0060860

[118] Jumelle C, Mauclair C, Houzet J, Bernard A, He Z, Forest F, Peoc'h M, Acquart S, Gain P, Thuret G. Delivery of molecules into human corneal endothelial cells by carbon nanoparticles activated by femtosecond laser. PLoS One. 2015;**10**:e0132023. DOI: 10.1371/journal.pone.0132023

[119] Ljubimov AV, Saghizadeh M. Progress in corneal wound healing. Progress in Retinal and Eye Research. 2015;**49**:17-45. DOI: 10.1016/j.preteyeres.2015.07.002

[120] Miller-Kleinhenz JM, Bozeman EN, Yang L. Targeted nanoparticles for image-guided treatment of triple negative breast cancer: Clinical significance and technological advances. Wiley Interdisciplinary Reviews: Nanomedicine and Nanobiotechnology. 2015;**7**:797-816. DOI: 10.1002/wnan.1343

[121] Kiessling F, Mertens ME, Grimm J, Lammers T. Nanoparticles for imaging: Top or flop? Radiology. 2014;**273**:10-28. DOI: 10.1148/radiol.14131520

[122] Lu ZX, Liu LT, Qi XR. Development of small interfering RNA delivery system using PEI-PEG-APRPG polymer for antiangiogenic vascular endothelial growth factor tumor-targeted therapy. International Journal of Nanomedicine. 2011;**6**:1661-1673. DOI: 10.2147/IJN.S22293

[123] Moysidis SN, Alvarez-Delfin K, Peschansky VJ, Salero E, Weisman AD, Bartakova A, Raffa GA, Merkhofer RM Jr, Kador KE, Kunzevitzky NJ, Goldberg JL. Magnetic field-guided cell delivery with nanoparticle-loaded human corneal endothelial cells. Nanomedicine. 2015;**11**:499-509. DOI: 10.1016/j.nano.2014.12.002

Section 2

Toxicity Validation for Nanomedicine

Chapter 4

Cytotoxic and Antiproliferative Effects of Nanomaterials on Cancer Cell Lines: A Review

Marcelo Grijalva, María José Vallejo-López,
Lizeth Salazar, Javier Camacho and Brajesh Kumar

Additional information is available at the end of the chapter

http://dx.doi.org/10.5772/intechopen.71685

Abstract

Cell models for the study of antiproliferative and/or cytotoxic properties of engineered nanoparticles are valuable tools in cancer research. Several techniques and methods are readily available for the study of nanoparticles' properties regarding selective toxicity and/or antiproliferative effects. Setting up of those techniques, however, needs to be carefully monitored. Harmonization of the wide range of methods available is necessary for assay comparison and replicability. Although individual or core laboratory capabilities play a role in selection and availability of techniques, data arising from cancer cell models are useful in guiding further research. The variety of cell lines available and the diversity of metabolic routes involved in cell responses make *in vitro* cell models suitable for the study of the biological effect of nanoparticles at the cell level and a valid approach for further *in vivo* and clinical studies. The present systematic review looks at the *in vitro* biological effects of different types of nanoparticles in cancer cell models.

Keywords: cancer, nanoparticles, organic, metallic, nanobiotechnology, cytotoxicity, antiproliferation

1. Introduction

Toxicity studies are needed for nanoparticles' (NPs) intended application on biomedical theranostics. Nanostructures are being designed and fabricated with a wide range of potentialities, including those in cancer therapeutics, medical imaging and diagnostics. Thus, research on cell models and *in vivo* toxicity is growing as the nanostructures that are being fabricated will find possible uses in biomedical, clinical medicine and health-related sectors. NPs have interesting physical-chemical properties that are of value when engineering drug delivery systems, diagnostic platforms and nanotechnology-based imaging strategies. The high surface-to-volume ratio of NPs allows the use of different molecules, such as those intended for targeted drug delivery [1]. The above properties, however, might render NPs a toxic *in vitro* and *in vivo* profile. Since many NPs are entering the health market, it becomes increasingly necessary to perform toxicological investigation along with NP fabrication.

Earlier and recent toxicity studies on human cell lines have found a range of nanostructures that might be selectively toxic for particular cellular lines, including cancerous ones [2, 3]. This selective toxicity against specific types of cancer is a promising research field with potential implications in (pro)diagnosis and therapeutics [4, 5]. Human cell models are available for a variety of malignancies, serving as suitable platforms for exploring antiproliferative and cytotoxic effects of nanostructures [6]. Data from cancer cell models and NP exposure are valuable for guiding and designing *in vivo* testing and, potentially, for developing new anticancer theranostic strategies [7].

In this review, we compile and discuss the findings of several recent works using cancer cell models and exploring selective NP toxicity and/or antiproliferative effects for potential therapeutic applications in cancer. We looked for particularly interesting scientific papers from indexed journals published within 2015–2017. The focus of this review is on methodological aspects of NP treatment on human cell–based models, i.e. viability assessment techniques, experimental design for investigation of mechanisms of cellular damage, cell culture protocols and NP stability assessment, including in biological media. Results of this review are presented by nature of NPs. Studies exploring new cell culture techniques for assessment of NP toxicity on cancer cell lines were also included.

2. Physicochemical characteristics of nanomaterials and their influence on toxicity

The potential for biomedical applications of several NPs is enormous. There are, however, several shortcomings regarding interactions of engineered NPs with biological environments. Toxicity concerns for NPs intended for use in biomedicine have limited their translation into clinical settings. NP properties such as size, surface-to-volume ratio, shape, surface functionalization and stability on biological media, among others, have been demonstrated to influence the toxicological profile of the nanostructures and their biocompatibility in general [7, 8]. It has been also demonstrated that the level of toxicity varies depending upon cell type, which reflects on particular cell line biology and genetics [9].

Interactions of NPs inside a biological environment, e.g. eukaryotic cells, have been widely studied [10]. Proteins, lipids or any biomolecule may be absorbed by NPs, affecting not only the original synthetic structure but its biological effect. Assessing antitumor properties of NPs requires stability in investigation under *in vitro* cell culture conditions. The interactions between NPs and biomolecules present in the culture media, such as proteins and lipids, could change nanomaterial's characteristics [12]. For instance, research has demonstrated the formation of protein corona around NP surface due to interaction with cellular media, resulting in modifications of their physical properties and leading, for instance, to aggregation and

Nanoparticles		Anticancer properties	
		Advantages	**Disadvantages**
Metallic and nonmetallic nanoparticles	Naked [94, 95]	• High antitumor activity • Storage and release of energy to other molecules quite effectively • Improvement of sensitive single-molecule detection techniques • External stimuli responsive, e.g. light and magnetism modulate its activity • Tunable physical and chemical properties	• Conformational changes • Coalescence • Stabilizers do not function properly in different solvents • In a large extent, synthesized with toxic chemicals for health and/ or environment
	Coated [13, 16, 96]	• Easy conjugation to drugs, proteins, and/or nucleotides • Attenuated cytotoxicity against normal cells due to surface functionalization • Specific site of action	• Biological effect varies among different coatings • Formation of a protein corona • Sedimentation and/or aggregation
Liposomes [96]		• High biocompatibility • Capability of conjugation with soluble and insoluble drugs • Targeted drug release • Low toxicity	• Colloidal stability and biodegradability • Complex and expensive synthesis

Table 1. Advantages and disadvantages of (non)metallic nanoparticles and liposomes application in cancer research.

sedimentation [13]. Thus, NP characterization during *in vitro* experiments is essential to understand the relationship between physical properties and mechanisms of *in vitro* toxicity.

In general, smaller NPs are more toxic than larger ones [14]. Several works have confirmed this relationship and some authors have identified NP sizes that correlate well with the level of toxicity observed on *in vitro* tests [15]. A range of toxic mechanisms leading to apoptosis, necrosis and genotoxicity is triggered by NPs of different range of dimensions. The net cytotoxic effect is usually cell and NP concentration dependent [9].

Several coating strategies have been tested for lowering the cytotoxic effects of many engineered NPs intended for medical applications. Metallic NPs have been extensively investigated and are excellent candidates as drug nanocarriers, for imaging strategies and for immunological platforms in biomedicine [11]. Toxicity concerns have, however, slowed their faster development and translation. Green chemistry or biologically mediated synthesis of coated metallic NPs is on the rise, and consequently their nanotoxicity evaluation on biological media has been pursued and published [15, 16].

Nanostructures such as semiconductor quantum dots (QDs) are also being investigated for biomedical purposes. Since the toxicity of these nanostructures is known, different coating procedures have been investigated in order to reduce their toxicity. For instance, zinc sulfide (ZnS) QDs functionalized with chitosan have shown no toxic effects on human leukocytes, contrary to the highly toxic cadmium sulfide (CdS) QDs that, even coated with biocompatible chitosan, showed to be toxic in a concentration and time-dependent manner [17]. A summary of the *pros* and *cons* of the use of NPs in cancer research is shown in **Table 1**.

3. The selective toxicity of nanomaterials on *in vitro* cancer cell models

Several mechanisms are involved in NP-mediated *in vitro* toxicity in normal (i.e. noncancerous) and cancerous cells. Cellular responses to NP exposure might include those at cell, organelle and gene level or a combination of them [18]. Direct cytotoxic effects might be apoptosis or necrosis (or both) mediated, with a number of mechanisms leading to cell death, changes in proliferation patterns and effects on cell differentiation. High levels of reactive oxygen species (ROS) production, downregulation of antioxidant enzyme coding genes, lipid peroxidation and genotoxic effects, among others, may be involved in the integrated cellular response to NPs [19, 20].

In spite of the number of studies providing useful information on nanotoxicological profiling, there remains particular information with regard to cell-NP specificity interactions. In addition, investigation on the toxicity of nanostructures and biointeractions rely on data from a wide variety of experiments with several different methods and techniques that are chosen on the basis of laboratory capabilities and researchers´ technical expertise [21, 22]. Then, there are, as today, no standard cell panels or defined protocols available for assessment of cancer cell responses to NPs; therefore, data arising from those studies are difficult to compile and integrate. Moreover, there is still the risk that the toxicological picture from a particular study on specific NPs and cell lines might not be "complete" enough and that toxic risks may be overlooked.

Apoptosis is a common response of cells to NP treatment. Azizi and colleagues found that albumin-coated silver NPs (AgNPs) LD50 were several times lower for breast cancer cells than for normal white blood cells. Apoptosis assays such as Annexin V and microscopy counts of apoptotic bodies demonstrated that albumin-coated AgNPs exert proapoptotic selective effects on breast cancer cells while normal blood cells remained viable at the tested concentrations and times of exposure [5].

In a recent work on several murine cancer cell lines, Namvar and colleagues investigated the antitumor properties of biosynthesized zinc oxide NPs (ZnONPs). They found that cancer cell proliferation was inhibited by NPs in a time- and concentration-dependent manner and that the mechanism of cell death was primarily apoptosis via procaspases activation and intrinsic mitochondrial pathway triggering [2].

NP exposure may cause cancer cell death by oxidative stress through varied mechanisms, including ROS production, inhibition of antioxidant enzymes, mitochondrial damage and lipid peroxidation [20]. For instance, Matulionyte, et al. demonstrated that photoluminescent gold nanoclusters have specific toxicity against MCF-7 breast cancer cells and were less toxic on MDA-MB 231 breast cancer cells, a highly drug-resistant cell line. The mechanism of cell death was apoptosis, necrosis and generation of ROS, effects that were more evident in MCF-7 cells [23].

Several other mechanisms are involved in the selective toxicity of NPs against different cancer cell lines. Endoplasmic reticulum (ER) autophagy is a well-known process related with NP exposure. A study by Wei, et al. found that silica NPs (SiNPs) induced ER autophagy in colon cancer cells. The authors showed a time-dependent effect of NP exposure, but interestingly, autophagy was present only at either low or high NP concentrations [24].

Due to the complexity of cell responses to NPs, it is important to evaluate the biological effect of NPs from different perspectives, from toxicology assessment to both *in vitro* and *in vivo* testing, to better understand NP-induced cellular responses and the mechanisms behind them (**Figure 1**).

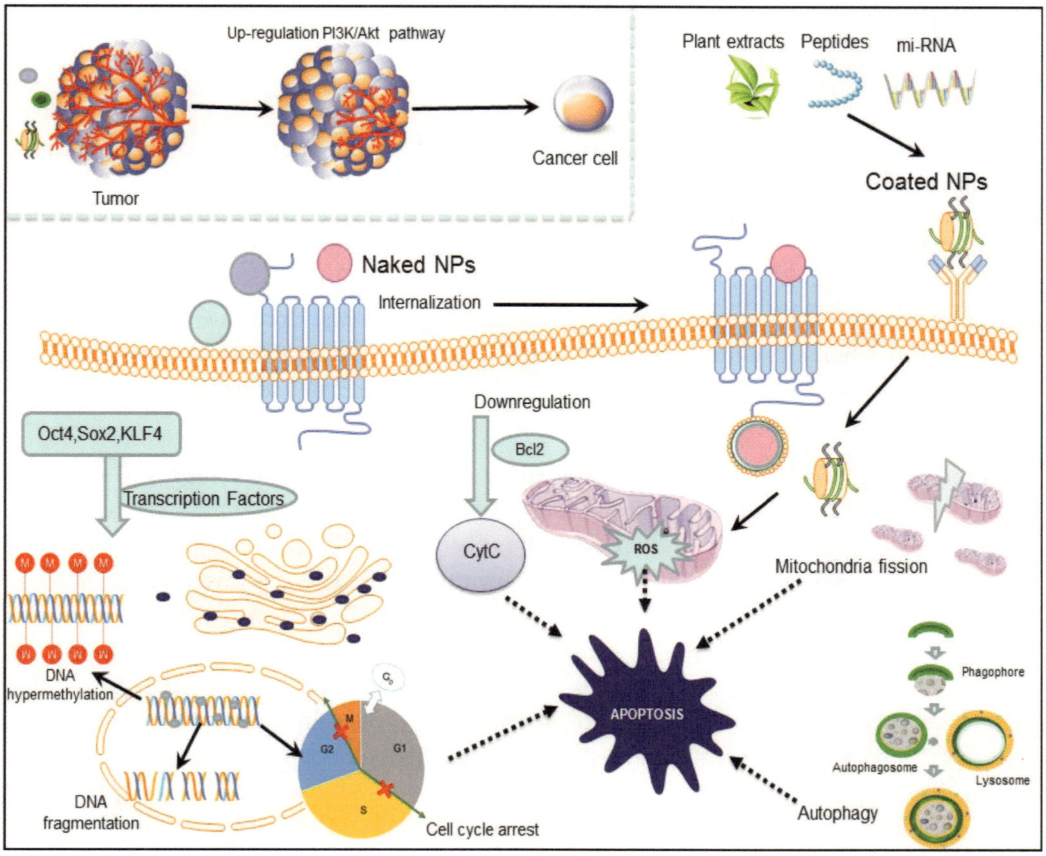

Figure 1. Schematic interpretation of nanoparticle (NP) cellular effects. NPs undergo internalization by nonspecific or specific endocytosis and remain in the cytoplasm or inside intracellular vesicles, either individually or in aggregates. NPs might release ions that enter the nucleus and cause DNA fragmentation/hypermethylation and/or cell cycle arrest in cancer cells. Furthermore, NPs' inhibitory effect on cellular viability is due to downregulation of antiapoptotic genes, e.g. Bcl2, generation of reactive oxygen species (ROS), mitochondria fission and autophagy and events that finally induce cell death through apoptosis. NPs could decrease the expression of transcription factors involved in stemness and thus inhibit angiogenesis.

In the following sections, we discuss the cytotoxic and antiproliferative *in vitro* properties of different types of NPs and their potential application in nanotherapeutics.

3.1. Metallic nanoparticles: noble metals and selective antitumor properties

Inorganic nanostructures exhibit interesting physical properties such as magnetism, fluorescence and localized surface plasmon resonance, which in combination with NPs' small dimensions make them suitable for biological applications. An advantage over other types of nanostructures is that inorganic NPs could respond to external stimulation with light or magnetic fields [1]. Among inorganic NPs, noble metals have been commonly used for the synthesis of nanomaterials. For instance, silver, gold and platinum NPs are of interest in cancer research as multifunctional anticancer agents due to their particular properties [25, 26]. In the subsequent sections, antitumor properties of noble metallic NPs are discussed focusing on their *in vitro* effects on several cancer cell lines.

3.1.1. Silver nanoparticles

Silver nanoparticles (AgNPs) possess particular physicochemical properties that determine their extent of cytotoxicity in biological systems [27]. It is well documented that AgNPs exert an antiproliferative effect on cancer cell lines [19, 28]. According to Choi, et al., AgNPs develop a potential cytotoxic effect on A2780 ovarian carcinoma cells and ovarian cancer stem cells (OvCSCs) at high concentrations. The inhibitory effect on cellular viability is caused by the upregulation of p53 and caspase-3 genes. In contrast, AgNPs might promote cell proliferation at low concentrations. The relevance of these findings is that OvCSCs present more sensitivity to the treatment with AgNPs, which is particularly interesting due to the fact that CSCs might increase the risk of acquired resistance to chemotherapy [19].

The therapeutic effect of AgNPs in multidrug resistant (MDR)-cancer cells has also been investigated. Kovacs, et al. demonstrated that AgNPs induce apoptosis-mediated cell death in drug-sensitive (Colo 205) and drug-resistant (Colo 320) colon adenocarcinoma cell lines, in a dose-dependent manner [28]. The internalization of AgNPs was observed in both cell types; thus, they remained in the cytoplasm. In addition, AgNPs may act synergistically with anticancer drugs to enhance their tumor-killing effects in MDR cells due to their capability of modulating efflux activity [28]. It is important to highlight the risk of exposing normal cells to AgNPs. To illustrate, a hippocampal neuronal cell model (HT22) was treated with AgNPs, obtaining a decrease in cell viability, oxidative damage and hypermethylation in DNA due to the internalization of AgNPs. These effects in normal cells may be prolonged since harmful impacts remain after AgNP removal [29]. Similar reports were found by Gao, et al., demonstrating that AgNPs can potentially damage mouse embryonic stem cells [30]. A novel approach to reduce cytotoxicity against normal cells is the functionalization or modification of AgNP surface [16]. Extensive research has been conducted to validate the hypothesis that AgNPs could inhibit angiogenesis, a complex process that is involved in the formation of new blood vessels and tumor progression [31]. For instance, Gurunathan, et al. concluded that the treatment of bovine retinal endothelial cells (BRECs) with AgNPs might activate PI3K/Akt pathway resulting in the inhibition of capillary formation [32]. Based on this evidence, AgNPs are potent antineoplastic agents with acute cytotoxic effects that modulate several metabolic pathways leading to decreased cell viability, independently or in combination with other anticancer drugs. This synergistic effect will be further discussed along this chapter.

3.1.2. Gold nanoparticles

Compatibility of gold with biosystems has been well demonstrated since metallic nanoscale materials were originally developed [33]. In recent years, synthesis and application of gold NPs (AuNPs) in the biomedical field have substantially increased due to their ductility physicochemical properties and biocompatibility. AuNPs can be synthesized in different shapes including spheres, rods, cubes, triangles, cones and shells [34]. Therefore, based on their size and shape, "naked" AuNPs possess several applications, e.g. as antitumor agents, drug nanocarriers, hyperthermia enhancers and radio sensitizers [1, 35].

AuNPs exert *in vitro* cytotoxicity on several human cancer cell lines including cervical (HeLa), prostate (PC-3), hepatocellular carcinoma (HepG2) and breast cancer (MDA-MB-231) [3, 36–38]. Wozniak, et al. proved that spherical and rod-shaped AuNPs are more efficient than other shapes in reducing cell proliferation of cancer cells *in vitro* [36]. Positively charged naked AuNPs interact with negatively charged cell membranes, increasing cellular uptake, preferably with smaller diameter particles rather than larger ones [37]. Also of interest, AuNPs can be used in combination with other anticancer molecules. For instance, Ke, et al. reported that AuNPs improved the responsiveness of Calu-1 epidermoid carcinoma cell line to tumor necrosis factor (TNF)-related apoptosis-inducing ligand (TRAIL) [39]. This combined approach induced DNA fragmentation, mitochondrial fission and a decrease in cell viability due to apoptosis. By contrast, the effect on cell viability was minimal in the BEAS-2B normal lung cell line [39].

In addition, AuNPs could act as enhancers of hyperthermia-targeted therapy because they efficiently absorb laser light and convert it into thermal energy [40]. The synergistic effect of AuNPs and laser-induced thermotherapy renders thermally exposed cancer cells susceptible to be ablated with minimal exposure times and lower laser intensities [33]. Rau, et al. showed that AuNPs could cause severe damage in the cytoskeleton of MG63 osteosarcoma cells in combination with laser treatment, increasing the calcium content inside the cells and leading to mineralization [41]. Another technique to induce hyperthermia in tumors is directed ultrasound. Kosheleva, et al. discovered that the combined treatment of ultrasound and AuNPs exerted a more acute cytotoxic effect on A549 lung cancer cells compared to BEAS-2B normal lung cells when cultivated separately and in coculture [42]. These findings suggested that AuNP-assisted thermotherapy could cause targeted cancer cell ablation, while avoiding damage to surrounding noncancerous cells.

AuNPs can be uptaken by cancer cells via endocytosis and trigger apoptotic events [43]. As a consequence, an improvement in radiation therapy has been observed when cancer cells are previously exposed to AuNPs [43]. Likewise, high atomic number in AuNPs increases radiation absorption from the target tumor [43]. Literature suggests that AuNPs act as radiosensitizers in several cancer cell lines, such as U251 glioblastoma, which in clinical practice could increase radiotherapy efficacy and prevent the development of drug-resistant tumors [44]. Another approach thoroughly studied by Rezaee, et al. showed that electroporation enhances radiosensitizing effect of AuNPs in HT-29 colon adenocarcinoma cells as a result of increasing cell membrane permeability. In this study, AuNPs' radiosensitizing effect was more prominent in cancer cells than in normal counterparts [43].

3.1.3. Platinum nanoparticles

Several investigations have addressed the antiproliferative effect of platinum nanoparticles (PtNPs) in cell models [45–48]. Bendale, et al. concluded that the harmful effect of PtNPs on cancer cell viability depends on the cell type. At the same PtNP concentration, an acute cytotoxic effect was observed in lung (A549), ovary (PA-1) and pancreatic (Mia-Pa-Ca-2) cancer cells [45] . In this study, no significative effect on cell viability was observed in breast, renal, colon and leukemia cancer cell lines. Interestingly, peripheral blood mononuclear cells (PBMCs) were not affected either, suggesting that PtNPs could preferably target tumor cells [45]. According to Kutwin, et al., PtNPs severely affect the proliferation rate and morphology of U118 and U87 human malignant glioma cell lines, and as a consequence, cells suffer from membrane disruption, reduced density and decreased migration [46]. Gehrke, et al. did not find any adverse effect on cellular viability when HT29 colon carcinoma cells were treated with PtNPs. It was observed, however, that smaller PtNPs enter the cells and remain in the cytoplasm or inside intracellular vesicles, either individually or in aggregates. Additionally, PtNPs released Pt ions that may bind to DNA leading to strand cleavage damage [49]. Another important feature is the synergistic antitumor activity between platinum and gold NPs. Ahamed, et al. reported that platinum-coated gold nanorods (AuNRs-Pt) affected cell viability on MCF7 breast cancer cells at relatively low doses. The mechanism of action of AuNRs-Pt involved impairment of normal morphology resulting in rounded cells, cell cycle detention at SubG1 phase, increased expression levels of proapoptotic genes caspase-3 and caspase-9 and generation of ROS [47]. Manikandan, et al. demonstrated that PtNPs could improve photothermal treatment in cancer cells. Neuro-2a brain neuroblastoma cells were exposed to the combined scheme of laser irradiation and PtNPs, which resulted in induction of apoptosis [48]. There was no significative effect on cellular viability when PtNPs and laser treatment were applied separately [48].

3.1.4. Other metal-based nanomaterials

Titanium dioxide (TiO_2), zinc (Zn), copper (Cu) and iron (Fe) are used in several industrial applications such as cosmetics, paint chemicals, food additives, pharmacological coatings, drug delivery systems, biosensor technologies and body implants. These nanomaterials have been also tested in cancer research and development of new therapeutics [22, 50].

Xia and coworkers reported the cytotoxic effect of cuprous oxide nanoparticles (CONPs) on HeLa, SiHa and MS751 human cervical cancer cell lines. Results demonstrated that CONPs are uptaken by cells and internalized into the cytoplasm, mitochondria and lysosomes; as a result, cell morphology alterations and decreased cellular viability were observed. Cell cycle arrest in the G1/G0 phase, induction of apoptosis and autophagy were also reported [51].

The antineoplastic effect of CONPs in PC-3, LNCaP FGC and DU145 human prostate carcinoma cells was investigated by Wang, et al. The results of this study suggest that CONPs might induce cytotoxicity selectively on cancerous cells without affecting normal prostate epithelial cells (RWPE-1). Moreover, a significant decrease in the expression of Oct4, Sox2 and KLF4 transcription factors related with stem-cell proliferation capability was observed [52].

Superparamagnetic iron oxide nanoparticles (SPIONs) are also included in a large extent in nanomedical products [1]. SPIONs develop magnetic properties within a magnetic field; therefore, they are able to act in specific target sites [1]. Several studies demonstrated that SPIONS can be approached as hyperthermia enhancers, contrast agents in magnetic resonance imaging, drug nanocarriers and anticancer candidates [1]. For instance, Du, et al. studied the combined effect of SPIONs and spinning magnetic field (SMF) on the survival rate of U-2 OS and Saos-2 osteosarcoma cell lines. This combined treatment exerts a more effective cytotoxic response triggering the intracellular ROS generation, autophagic cell death and apoptosis, than SPION treatment alone [53].

3.2. Nonmetallic and organically coated metallic nanomaterials: antiproliferative and cytotoxic properties

3.2.1. Green synthesis–based nanomaterials

Production of materials at the nanometric scale (1–100 nm) has been performed using several approaches [54]. The most common synthesis method involves the use of three elements: capping agent, reducing agent and solvent [54]. However, most of these elements are toxic, flammable, corrosive and even dangerous for the natural environment and living organisms. For this reason, a new green chemistry tendency emerged in the nanotechnology area to modify chemical processes and reduce or minimize the use of hazardous reagents [55]. The green-synthesis approach has been focused on finding nontoxic elements to develop a more eco-friendly design with improved efficiency [54]. Some of these new techniques require the use of solvents such as water, supercritical CO_2 or ionic liquids [56, 57]. For example, silver and gold nanoscale structures, due to their chemical and biological properties, have been widely used in green synthesis in combination with medicinal plants (photosynthesis) or bacterial/fungi/viral proteins (microbial-synthesis) [58]. This section provides further interesting examples of green-synthetized nanomaterials.

3.2.1.1. Photosynthesis

The importance of developing an alternative nanosynthesis protocol is not only for an environmental footprint reduction, but contributes also for the simplification of industrial production with the lack of expensive organic solvents and toxic chemicals [59]. The use of innocuous plant extracts with solvents such as water facilitates the production and further evaluation of green nanomaterials, which are fundamental for biological applications in critical areas e.g. drug production [59]. There are several nanoscale structures coated with plant extracts and their effect on living systems has been extensively studied [60]. For instance, Krishnaraj, et al. reported that Ag/Au biosynthesized NPs with *Acalypha indica* extract exerted a cytotoxic effect in MDA-MB-231 human breast cancer cells. These NPs exhibited a proapoptotic effect through caspase-3 activation [58]. Another example of naturally coated AgNPs includes the effect of the *Erythrina indica* extract causing a dose-dependent reduction of viability in MCF7 breast cancer cells and HepG2 hepatocellular carcinoma [27]. The authors also demonstrated high antimicrobial activity for AgNPs against *Staphylococcus aureus, Micrococcus luteus, Escherichia coli, Bacillus subtilis, Salmonella typhi* and *Salmonella paratyphi* [27]. Moreover, AgNPs were synthesized using *Albizia adianthifolia* leaf extract; the AgNP analysis determined the presence of saponins and glycosides as stabilizing agents [61]. Toxicity analysis was performed on A549 lung cancer cells and normal peripheral lymphocytes [61]. The results showed a reduction in A549 cellular viability to 21% at 10 g/mL and 73% at 50 g/mL after 6 h of exposure [61]. In comparison, proliferation rates for normal cell lines were not altered [61]. Other applications of these nanostructured particles for disease treatment include antidiabetic effects, described with *Cassia fistula* AuNPs that reduce glucose levels in rats with streptozotocin-induced diabetes [62], and antimosquito larvicidal activity of *Nelumbo nucifera* AuNPs [63].

3.2.1.2. Microbial synthesis

Bacterial survival in the presence of heavy metals is caused by a transformation (reduction/precipitation) of metal ions into insoluble nontoxic metal nanoclusters. These detoxification reactions are mediated by intracellular accumulation or a physicochemical process–denominated extracellular biosorption, which facilitates the concentration of contaminants, e.g. heavy metals, and binds them in their cellular structure, with variable levels of dispersity [64]. Based on these bacterial properties, Klaus, et al. described AgNP production in *Pseudomonas stutzeri*. This bacterium reduces silver ion to generate Ag^0 and AgS_2 NPs of different shapes and sizes located around the cellular poles [65]. Another interesting example is *B. subtilis* that reduces Au^{3+} to a neutral nanocompound (Au^0) [66]. Moreover, production of lipopolysaccharides and phospholipids in some bacteria mediates bioreduction, e.g. transformation of chloroauric acid ($HAuCl_4$) to AuNPs in *E. coli* DH5α [64]. Nonetheless, assembly of microbial NPs is also performed in several fungi species such as *Penicillium chrysogenum*, which has showed to be an AuNP producer in $HAuCl_4$ solution [67]. Another remarkable study optimized nanowire production with M13 virus as biotemplate for development of lithium batteries [68]. Based on this information, affordable and massive industrial production should be feasible with the use of biological nanofactories such as the above-mentioned examples. However, the lack of complete understanding of the molecular reaction mechanism is a major disadvantage of this methodology.

3.2.2. Organically coated metallic and nonmetallic nanomaterials

Nanobiotechnology as a mature biomedical field emerged in the last years [69]; for example, from gene-delivery systems to targeted drug delivery, it has several applications in cancer treatment, diagnosis (biomarkers), molecular biology and genetic/cell engineering [70, 71]. A nanomedicine-based therapeutic approach might be built on nanocarriers, e.g. liposomes and NPs that improve chemotherapeutic biodistribution [72] and have been useful for treating diseases such as cancer [73] and microbial infections [74]. In 1989, Matsumura and Maeda described the enhanced permeation and retention (EPR) effect, a controversial concept based on the passive accumulation of macromolecular drugs in tumors due to the presence of a high number of abnormal blood vessels (angiogenesis), which lack lymphatic drainage, affecting in turn as a drug delivery system [75]. Despite the fact that this effect has been extensively studied but has failed in clinical trials [76], EPR is still one of the most used concepts in nano-biodistribution [76]. With this information in mind, this section discusses relevant aspects of metallic and nonmetallic coated nanomaterials, including liposomes as novel therapeutic agents for cancer.

3.2.2.1. Organically coated nanomaterials

Organic coating is used to stabilize NPs and maintain a balance between electrostatic and electrosteric repulsion forces [73]. NPs of different shapes might be covered with diverse capping agents such as citric acid, polysaccharides, surfactants, proteins, polymers and nucleic acids [77, 78]. However, despite the fact that they have the same core material, coated-NPs exert different biological responses. For instance, viability, genotoxicity and mutagenicity evaluation of AgNPs coated with anionic (citrate, SDS), neutral (disperbyk, tween) or cationic (byk and chitosan) compounds were performed by Kun, et al. using lymphoblast TK6 cell line and Chinese hamster lung fibroblasts. The methodology used for testing involved trypan blue exclusion assay, relative growth activity, cell morphology, HPRT mutation and comet assay. The results determined that AgNPs_byk and AgNPs_chitosan were the most cytotoxic, affecting cell morphology, inhibiting proliferation and inducing cell death through apoptosis or necrosis. Furthermore, AgNPs_byk showed significant mutagenic effects by inducing DNA strand breaks and oxidation. It is important to note that AgNPs_byk formed the smallest agglomerates in medium solution in comparison to other coated AgNPs, suggesting that size is an important factor in toxicity. To sum up, coated NPs display various biological *in vitro* effects depending mainly on their surface charge [12].

One of the well-known NP biointeractions is that with bovine serum albumin (BSA), which relies on principle that a protein corona is dynamically formed around NPs when they enter a biological environment [79]. A recent investigation performed by Zhou, et al. determined that ZnONPs bound to BSA elicited interleukin-6 (IL-6) production-mediated anti-inflammatory responses in HepG2 liver cancer cell line. Additionally, synthesized NPs induced mitochondria and lysosomal damage by increasing intracellular Zn ions production [79]. However, the analysis of the biological effect of ZnONPs bound to α-linolenic acid (LNA) did not show the same response [79]. Another interesting example is the evaluation of the response of SPIONs conjugated to the antitumor peptide ATWLPPR (A7R) on HUVEC human umbilical vein endothelial cells and MDA-MB-231 human breast adenocarcinoma [11]. These NPs might be adjusted carriers for targeted drug delivery systems using their magnetic properties. Furthermore, the presence of cell receptors for the A7R peptide facilitates the uptake of the nanocarriers. This is particularly important because the role of the receptor is to repress the vascular endothelial growth factor A (VEGF A). Consequently, NPs affect angiogenic events and impair cell proliferation [11].

The study of commercial anticancer drug formulations in nanoform has also been evaluated, with positive results in many cases. For instance, tamoxifen, an anticancer agent used in estrogen receptor-positive (ER+) breast cancer, has been commonly used before surgery to reduce tumor volume [80]. However, tamoxifen resistance has become a significant problem in cancer treatment [80]. Devulapally, et al. synthesized biodegradable polymer NPs loaded with the active compound of tamoxifen (4-hydroxytamoxifen-4OHT) and the noncoding RNA (anti-miR-21). NPs showed antiproliferative and proapoptotic effects in human breast (MCF7, ZR-751, BT-474) and mouse mammary (4 T1) carcinoma cells [81].

3.2.2.2. Nonmetallic nanomaterials

Despite the fact that most widely used nanomaterials have metal cores, a number of industry-relevant nonmetallic nanoscale particles such as SiO_2 and carbon NPs have been engineered. For instance, silica NPs have been extensively used in food additives [82], toothpaste and skin care products [83]. However, their use requires toxicology screening to determine their innocuity. According to Wittig, et al., commercially available nanosilica (Ø 12 nm) increases the growth of GXF251L human gastric carcinoma cells. The results showed an important proliferative effect through the activation of cellular epidermal growth factor receptor (EGFR) and mitogen-activated protein kinase (MAPK) signaling pathways [84].

On the contrary, research on antiproliferative properties have found that cerium oxide nanocrystals (nanoceria: CeO_2-NCs) can act as an anticancer drug [85]. The investigation conducted by Khan, et al. found that fluorescence microscopy assessments of nanoceria displayed a marked *in vitro* cytotoxic effect and reduced cellular viability on HT-29 human colorectal adenocarcinoma. The results showed downregulation of Bcl2 and BclxL protein expression suggesting proapoptotic effects. Additionally, this study confirmed previous reports [86] where cerium oxide exhibited a cytotoxic effect toward cancer cells with minimum toxicity to normal cells [86]. Another interesting example involved the evaluation of cytotoxic effects of smart-releasing NPs synthesized using cytochrome C (Cyt C) and hyaluronic acid (HA) [87]. This study, by Figueroa, et al., showed that A549 human lung adenocarcinoma cellular viability was reduced to 20% (0.16 mg/mL [Cyt C], 6 h of exposure), while OS-7 African green monkey kidney fibroblasts were not affected. Confocal microscopy imaging confirmed the release of Cyt C to the cytoplasm upon reaching the target. In this study, EPR effect is used to develop a new potential stimuli-driven nanoparticle for cancer treatment [87].

Liposomes were firstly described in the middle 1960s as spherical vesicles constituted with phospholipid bilayers [88]. These lipid-based nanoparticles have been used in several fields from biophysics to biology for many years [88]. With the advances of nanotechnology, liposomes have evolved in order to assure controlled delivery of active molecules to a specific site of action. For instance, a radiation therapy scheme in use for more than 50 years is boron neutron capture (BNCT), which is based on the specific delivery of the isotope (boron-10) undergoing a nuclear reaction to form boron-11, through exposure to a laser beam (neutron source [89]). This reaction causes a release of an α-particle that has a high linear energy transfer (LET) and kills the equivalent of one cell diameter [89]. In the above research, conducted by Maitz, et al., the effect of unilamellar liposomes (composed of cholesterol, 1, 2-distearoyl-sn-glycero-3-phosphocholine, $K[nido\text{-}7\text{-}CH_3(CH_2)15\text{-}7,8\text{-}C_2B_9H_{11}]$ and core $Na_3[1\text{-}(2'\text{-}B_{10}H_9)\text{-}2\text{-}NH_3B_{10}H_8]$) on mice bearing tumors (breast carcinoma EMT6 and colon carcinoma CT26) was studied.

http://dx.doi.org/10.5772/intechopen.71685

Type	Nanoparticles	Coating material	Cell type	Biological effect	Ref.
Metallic	AgNPs	Naked	A2780 ovarian carcinoma	Cytotoxic	[19]
			OvCSCs ovarian cancer stem cells		
			Colo 205 colon adenocarcinoma	Proapoptotic, synergic with anticancer drugs	[28]
			Colo 320 drug-resistant colon adenocarcinoma		
			HT22 hippocampal neuronal model	Antiproliferative, DNA hypermethylation and oxidative stress damage	[29]
			Mouse embryonic stem cells	Transcriptomic alterations	[30]
			Bovine retinal endothelia	Angiogenesis inhibition	[32]
		Erythrina indica extract	MCF7 breast cancer	Cytotoxic	[27]
			HepG2 hepatocellular carcinoma		
		Albizia adianthifolia extract	A549 lung cancer		[61]
		Acalypha indica extract	MDA-MB-231 human breast adenocarcinoma	Cytotoxic, proapoptotic	[58]
	AuNPs	Naked	HeLa cervical carcinoma	Cytotoxic	[3, 36–38]
			PC-3 prostate cancer		
			HepG2 hepatocellular carcinoma		
			MDA-MB-231 human breast adenocarcinoma		
			Calu-1 epidermoid carcinoma	Proapoptotic, synergistic effect with anticancer molecules, DNA fragmentation and mitochondrial fission	[39]
			MG63 osteosarcoma	Cytoskeleton damage in combination with laser treatment	[41]
			A549 lung cancer	Cytoskeleton damage in combination with ultrasound	[42]
			U251 glioblastoma	Proapoptotic, radiosensitizer	[43, 44]
			HT-29 colon adenocarcinoma		
		Platinum coated	MCF7 breast cancer	Proapoptotic, ROS production and cell cycle arrest	[47]
		Acalypha indica extract	MDA-MB-231 human breast adenocarcinoma	Cytotoxic, proapoptotic	[58]
	PtNPs	Naked	A549 lung cancer	Cytotoxic	[45]
			PA-1 human ovarian teratocarcinoma		
			Mia-Pa-Ca-2 human pancreas carcinoma		
			U118/U87 human malignant glioma	Antiproliferative	[46]
			Neuro-2a brain neuroblastoma	Proapoptotic in combination with laser treatment	[48]

Type	Nanoparticles	Coating material	Cell type	Biological effect	Ref.
	Cuprous oxide CONPs	Naked	SiHa cervical squamous carcinoma	Cytotoxic, autophagy and proapoptotic	[51]
			HeLa cervical carcinoma		
			MS751 cervical cancer		
			LNCaP FGC human prostate carcinoma	Cytotoxic, reduction in transcription factors for proliferation	[52]
			PC-3 prostate cancer		
			DU145 prostate carcinoma		
Metallic	Iron oxide, SPIONs	Naked	U-2 OS osteosarcoma	Cytotoxic, autophagy and proapoptotic	[53]
			Saos-2 osteosarcoma		
		antiangiogenic peptide ATWLPPR (A7R)	MDA-MB-231 human breast adenocarcinoma	Reduce angiogenesis and proliferation	[11]
	ZnO	Bovine serum albumin	HepG2 hepatocellular carcinoma	Anti-inflammatory and mitochondria-lysosome damage inducer	[79]
	Biopolymer	4-hydroxytamoxifen and noncoding RNA (anti-miR-21)	MCF7, ZR-751, BT-474 breast cancer	Antiproliferative and proapoptotic	[81]
			4 T1 mouse mammary carcinoma		
	SiO_2	Naked Ø 12 nm	GXF251L gastric carcinoma	Increase proliferation	[84]
Nonmetallic	Cerium oxide CeO	Naked	HT29 human colorectal adenocarcinoma	Cytotoxic, antiproliferative and proapoptotic	[85]
	Cytochrome C (Cyt C) and hyaluronic acid (HA)	Naked	A549 lung cancer	Antiproliferative	[87]
	Core: Na3 [1-(2′-B10H9)-2-NH3B10H8]	Cholesterol, 1, 2-distearoyl-sn-glycero-3-phosphocholine, K [nido-7-CH_3 (CH_2)15–7, 8-$C_2B_9H_{11}$]	EMT6 breast carcinoma	Increase radiosensitivity of tumors	[89]
			CT26 colon carcinoma		
Liposomes	Lipid-core	Poly(ε-caprolactone), capric/caprylic triglyceride, sorbitan monostearate and polysorbate 80	SK-Mel-28 human melanoma	Cytotoxic, proapoptotic and cell cycle arrest	[92]

Table 2. Biological effects of different types of nanoparticles (NPs) on cancer cells.

The results showed a 50% tumor reduction after 45 min of radiation, despite lower boron concentrations inside EMT6 tumor, in comparison to CT26. The average time for tumor growth, set as three times the pretreatment volume, was 38 days for BNCT-treated mice in comparison to 4 days for untreated controls. In conclusion, the authors found that liposomes were useful elements for increasing inherent radiosensitivity in selected tumors [89].

The use of liposomes has also been found useful for drug delivery systems [90]. Sadhu, et al. evaluated the cytotoxicity and antiproliferative effects of liposomes designed to increase the intracellular glutathione disulfide (GSSG) on B16 murine metastatic melanoma tumor cells (B16F10), human metastatic lung carcinoma cells (NCI-H226) and *in vivo* on C57BL/6 mice. Glutathione (GSH) is fundamental in the antioxidant defense against ROS [91]. Oxidation of GSH is mediated by a sulfhydryl residue from oxidative species and results in GSSG [91]. Analysis of GSSG has been a challenge since it is not inducible and neither cell membrane permeant. The results showed an important effect in the apoptotic pathway affecting cell migration, invasion and adhesion. Dacarbazine (the treatment option for melanoma) and GSSG liposomes showed a significant *in vivo* reduction of tumor proliferation (90% and 85%, respectively) [91].

Drewes, et al. demonstrated that lipid-core nanocapsules containing poly(ε-caprolactone), capric/caprylic triglyceride, sorbitan monostearate and polysorbate 80 affected cell proliferation and triggered cell cycle arrest on SK-Mel-28 human melanoma cells. Furthermore, nanocapsules induced apoptosis and necrosis on a murine model B16F10 (H2b) bearing B16 melanoma cell line [92]. To sum up, GSSG liposomes and lipid-core nanocapsules are potentially useful for antimetastatic treatment and as drug delivery systems for melanoma treatment, respectively [91]. Organically coated nanostructures, including liposomes, might exert antiproliferative cytotoxic properties against cancer cells/tumors but may also induce cell proliferation depending on the type of tumor and nanostructure used.

The wide spectrum of known cancer cellular responses to nanomaterials is summarized in **Table 2**.

4. Conclusion

In vitro cellular models for the study of antiproliferative and/or cytotoxic properties of engineered nanomaterials are valuable tools in cancer research. Cancer cell lines represent very easy-to-use models where different codelivery treatments might be tested. For instance, including chemotherapeutic drugs, siRNAs and antibodies in the same NPs should help lower drug concentrations and side effects as well as improve the therapeutic effect. Taking advantage of this type of approach in cancer cell lines might be of value when testing NPs in personalized medicine applications, when tumor cells from the patients are collected and either cultured or injected into *in vivo* vertebrate models. Recently, Rita, et al. reported the use of zebrafish xenotransplants [93] using colon cancer cell lines, SW480, SW620, and HT29, HCT116 and Hke3. Larvae were injected with cancer cells to develop mono-/polyclonal tumors, which were treated with different antiproliferation drugs. Results displayed differential drug sensitivities and support the potential application of this assay in personalized medicine and diagnostics. Such approaches should also decrease multidrug resistance rates.

Several techniques and methods are readily available for investigation of nanostructured particle properties regarding their selective cytotoxicity and/or antiproliferative effects. Setting up of those techniques, however, needs to be carefully monitored. Harmonization of the wide range of methods available is necessary for assay comparison and replicability.

To sum up, extended cell-based testing (*in vivo*) is necessary to obtain a complete understanding of the *in vitro* results. Although individual or core laboratory capabilities play a role in selection and availability of techniques, data arising from cancer cell models have demonstrated usefulness in guiding further research.

Author details

Marcelo Grijalva[1,2]*, María José Vallejo-López[1], Lizeth Salazar[2], Javier Camacho[3] and Brajesh Kumar[4]

*Address all correspondence to:rmgrijalva@espe.edu.ec

1 Departamento de Ciencias de la Vida y la Agricultura, Universidad de las Fuerzas Armadas ESPE, Sangolquí, Ecuador

2 Centro de Nanociencia y Nanotecnología, Universidad de las Fuerzas Armadas ESPE, Sangolquí, Ecuador

3 Departamento de Farmacología, Centro de Investigación y de Estudios Avanzados del Instituto Politécnico Nacional, México, D.F., México

4 Department of Chemistry, Kolhan University, Tata College, West Singhbhum, Jharkhand, India

References

[1] Henriksen-Lacey M, Carregal-Romero S, Liz-Marzán LM. Current challenges toward in vitro cellular validation of inorganic nanoparticles. Bioconjugate Chemistry. Jan. 2017;**28**(1):212-221

[2] Namvar F, et al. Cytotoxic effects of biosynthesized zinc oxide nanoparticles on murine cell lines. Evidence-Based Complementary and Alternative Medicine. 2015;**2015**:593014

[3] Moses SL, Edwards VM. Cytotoxicity in MCF-7 and MDA-MB-231 breast cancer cells, without harming MCF-10A healthy cells. Journal of Nanomedicine & Nanotechnology. 2016;**7**(2):369

[4] Ortega FG, et al. Study of antitumor activity in breast cell lines using silver nanoparticles produced by yeast. International Journal of Nanomedicine. 2015;**10**:2021-2031

[5] Azizi M, Ghourchian H, Yazdian F, Bagherifam S, Bekhradnia S, Nyström B. Anticancerous effect of albumin coated silver nanoparticles on MDA-MB 231 human breast cancer cell line. Scientific Reports. Jul. 2017;**7**(1):5178

[6] Chang J-S, Kuo H-P, Chang KLB, Kong Z-L. Apoptosis of hepatocellular carcinoma cells induced by nanoencapsulated polysaccharides extracted from *Antrodia camphorata*. PLoS One. 2015;**10**(9):e0136782

[7] Liu G, et al. Cytotoxicity of various types of gold-mesoporous silica nanoparticles in human breast cancer cells. International Journal of Nanomedicine. 2015;**10**:6075-6087

[8] Saifullah B, Hussein MZB. Inorganic nanolayers: Structure, preparation, and biomedical applications. International Journal of Nanomedicine. 2015;**10**:5609-5633

[9] Joris F, et al. The impact of species and cell type on the nanosafety profile of iron oxide nanoparticles in neural cells. Journal of Nanobiotechnology. Sep. 2016;**14**(1):69

[10] Foglia S, et al. In vitro biocompatibility study of sub-5 nm silica-coated magnetic iron oxide fluorescent nanoparticles for potential biomedical application. Scientific Reports. Apr. 2017;**7**:46513

[11] Niescioruk A, et al. Physicochemical properties and in vitro cytotoxicity of iron oxide-based nanoparticles modified with antiangiogenic and antitumor peptide A7R. Journal of Nanoparticle Research. 2017;**19**(5):160

[12] Huk A, et al. Impact of nanosilver on various DNA lesions and HPRT gene mutations—Effects of charge and surface coating. Particle and Fibre Toxicology. Jul. 2015;**12**:25

[13] Shannahan JH, Lai X, Ke PC, Podila R, Brown JM, Witzmann FA. Silver nanoparticle protein corona composition in cell culture media. PLoS One. Sep. 2013;**8**(9):e74001

[14] Joksić G, Stašić J, Filipović J, Šobot AV, Trtica M. Size of silver nanoparticles determines proliferation ability of human circulating lymphocytes in vitro. Toxicology Letters. Apr. 2016;**247**:29-34

[15] Guo X, et al. Size- and coating-dependent cytotoxicity and genotoxicity of silver nanoparticles evaluated using *in vitro* standard assays. Nanotoxicology. Oct. 2016;**10**(9):1373-1384

[16] Das B, et al. Surface modification minimizes the toxicity of silver nanoparticles: An in vitro and in vivo study. Journal of Biological Inorganic Chemistry. Aug. 2017;**22**(6):893-918

[17] Mansur AA, Mansur HS, de Carvalho SM, Lobato ZI, Guedes MI, Leite MF. Surface biofunctionalized CdS and ZnS quantum dot nanoconjugates for nanomedicine and oncology: To be or not to be nanotoxic? International Journal of Nanomedicine. 2016;**11**:4669-4690

[18] Patil NA, Gade WN, Deobagkar DD. Epigenetic modulation upon exposure of lung fibroblasts to TiO_2 and ZnO nanoparticles: Alterations in DNA methylation. International Journal of Nanomedicine. 2016;**11**:4509-4519

[19] Choi Y-J, et al. Differential cytotoxic potential of silver nanoparticles in human ovarian cancer cells and ovarian cancer stem cells. International Journal of Molecular Sciences. Dec. 2016;**17**(12):2077

[20] Abdal Dayem A, et al. The role of reactive oxygen species (ROS) in the biological activities of metallic nanoparticles. International Journal of Molecular Sciences. Jan. 2017;**18**(1):120

[21] Yang F, et al. Real-time, label-free monitoring of cell viability based on cell adhesion measurements with an atomic force microscope. Journal of Nanobiotechnology. Mar. 2017;**15**(1):23

[22] Kuku G, Culha M. Investigating the origins of toxic response in TiO_2 nanoparticle-treated cells. Nanomaterials. 2017;**7**(4):83

[23] Matulionyte M, Dapkute D, Budenaite L, Jarockyte G, Rotomskis R. Photoluminescent gold nanoclusters in cancer cells: Cellular uptake, toxicity, and generation of reactive oxygen species. International Journal of Molecular Sciences. Feb. 2017;**18**(2):378

[24] Wei F, Wang Y, Luo Z, Li Y, Duan Y. New findings of silica nanoparticles induced ER autophagy in human colon cancer cell. Scientific Reports. Feb. 2017;**7**:42591

[25] Rai M, Ingle AP, Birla S, Yadav A, Dos Santos CA. Strategic role of selected noble metal nanoparticles in medicine. Critical Reviews in Microbiology. Sep. 2016;**42**(5):696-719

[26] Fekrazad R, Naghdi N, Nokhbatolfoghahaei H, Bagheri H. The combination of laser therapy and metal nanoparticles in cancer treatment originated from epithelial tissues: A literature review. Journal of Lasers in Medical Science. Mar. 2016;**7**(2):62-75

[27] Vimbela GV, Ngo SM, Fraze C, Yang L, Stout DA. Antibacterial properties and toxicity from metallic nanomaterials. International Journal of Nanomedicine. 2017;**12**:3941-3965

[28] Kovacs D, et al. Silver nanoparticles modulate ABC transporter activity and enhance chemotherapy in multidrug resistant cancer. Nanomedicine. Apr. 2016;**12**(3):601-610

[29] Mytych J, Zebrowski J, Lewinska A, Wnuk M. Prolonged effects of silver nanoparticles on p53/p21 pathway-mediated proliferation, DNA damage response, and methylation parameters in HT22 hippocampal neuronal cells. Molecular Neurobiology. Mar. 2017;**54**(2):1285-1300

[30] Gao X, Topping VD, Keltner Z, Sprando RL, Yourick JJ. Toxicity of nano- and ionic silver to embryonic stem cells: A comparative toxicogenomic study. Journal of Nanobiotechnology. Apr. 2017;**15**(1):31

[31] Mukherjee S, Patra CR. Therapeutic application of anti-angiogenic nanomaterials in cancers. Nanoscale. Jul. 2016;**8**(25):12444-12470

[32] Gurunathan S, Lee K-J, Kalishwaralal K, Sheikpranbabu S, Vaidyanathan R, Eom SH. Anti-angiogenic properties of silver nanoparticles. Biomaterials. Oct. 2009;**30**(31):6341-6350

[33] Chen C-H, Chan T-M, Wu Y-J, Chen J-J. Review: Application of nanoparticles in urothelial cancer of the urinary bladder. Journal of Medical and Biological Engineering. 2015;**35**(4):419-427

[34] Amendola V, Pilot R, Frasconi M, Marago OM, Iati MA. Surface plasmon resonance in gold nanoparticles: A review. Journal of Physics Condensed Matter. May 2017;**29**(20):203002

[35] Orlando A, et al. Evaluation of gold nanoparticles biocompatibility: A multiparametric study on cultured endothelial cells and macrophages. Journal of Nanoparticle Research. Feb. 2016;**18**(3):58

[36] Wozniak A, et al. Size and shape-dependent cytotoxicity profile of gold nanoparticles for biomedical applications. Journal of Materials Science. Materials in Medicine. Jun. 2017;**28**(6):92

[37] Malugin A, Ghandehari H. Cellular uptake and toxicity of gold nanoparticles in prostate cancer cells: A comparative study of rods and spheres. Journal of Applied Toxicology. Apr. 2010;**30**(3):212-217

[38] Paino IMM, Marangoni VS, de Oliveira R d CS, Antunes LMG, Zucolotto V. Cyto and genotoxicity of gold nanoparticles in human hepatocellular carcinoma and peripheral blood mononuclear cells. Toxicology Letters. 2012;**215**(2):119-125

[39] Ke S, et al. Gold nanoparticles enhance TRAIL sensitivity through Drp1-mediated apoptotic and autophagic mitochondrial fission in NSCLC cells. International Journal of Nanomedicine. 2017;**12**:2531-2551

[40] Popp MK, Oubou I, Shepherd C, Nager Z, Anderson C, Pagliaro L. Photothermal therapy using gold nanorods and near-infrared light in a murine melanoma model increases survival and decreases tumor volume. Journal of Nanomaterials. 2014;**2014** Article ID 450670, 8 pages

[41] Rau L-R, Huang W-Y, Liaw J-W, Tsai S-W. Photothermal effects of laser-activated surface plasmonic gold nanoparticles on the apoptosis and osteogenesis of osteoblast-like cells. International Journal of Nanomedicine. 2016;**11**:3461-3473

[42] Kosheleva OK, Lai T-C, Chen NG, Hsiao M, Chen C-H. Selective killing of cancer cells by nanoparticle-assisted ultrasound. Journal of Nanobiotechnology. Jun. 2016;**14**(1):46

[43] Rezaee Z, Yadollahpour A, Bayati V, Negad Dehbashi F. Gold nanoparticles and electroporation impose both separate and synergistic radiosensitizing effects in HT-29 tumor cells: An in vitro study. International Journal of Nanomedicine. 2017;**12**:1431-1439

[44] Liu P, et al. Silver nanoparticles outperform gold nanoparticles in radiosensitizing U251 cells in vitro and in an intracranial mouse model of glioma. International Journal of Nanomedicine. 2016;**11**:5003-5014

[45] Bendale Y, Bendale V, Paul S. Evaluation of cytotoxic activity of platinum nanoparticles against normal and cancer cells and its anticancer potential through induction of apoptosis. Integrative Medicine Research. Jun. 2017;**6**(2):141-148

[46] Kutwin M, et al. Assessment of the proliferation status of glioblastoma cell and tumour tissue after nanoplatinum treatment. PLoS One. 2017;**12**(5):e0178277

[47] Ahamed M, Akhtar MJ, Khan MAM, Alhadlaq HA, Alrokayan SA. Cytotoxic response of platinum-coated gold nanorods in human breast cancer cells at very low exposure levels. Environmental Toxicology. Nov. 2016;**31**(11):1344-1356

[48] Manikandan M, Hasan N, Wu H-F. Platinum nanoparticles for the photothermal treatment of Neuro 2A cancer cells. Biomaterials. Jul. 2013;**34**(23):5833-5842

[49] Gehrke H, et al. Platinum nanoparticles and their cellular uptake and DNA platination at non-cytotoxic concentrations. Archives of Toxicology. Jul. 2011;**85**(7):799-812

[50] Schrand AM, Rahman MF, Hussain SM, Schlager JJ, Smith DA, Syed AF. Metal-based nanoparticles and their toxicity assessment. Wiley Interdisciplinary Reviews: Nanomed icine and Nanobiotechnology. 2010;**2**(5):544-568

[51] Xia L, et al. Cuprous oxide nanoparticles inhibit the growth of cervical carcinoma by inducing autophagy. Oncotarget. May 2017;**8**(37):61083-61092

[52] Wang Y, et al. Cuprous oxide nanoparticles inhibit prostate cancer by attenuating the stemness of cancer cells via inhibition of the Wnt signaling pathway. International Journal of Nanomedicine. 2017;**12**:2569-2579

[53] Du S, Li J, Du C, Huang Z, Chen G, Yan W. Overendocytosis of superparamagnetic iron oxide particles increases apoptosis and triggers autophagic cell death in human osteosarcoma cell under a spinning magnetic field. Oncotarget. Feb. 2017;**8**(6):9410-9424

[54] Fedlheim DL, Foss CA. Metal Nanoparticles: Synthesis, Characterization, and Applications. Marcel Dekker Inc. New York; 2002

[55] Poinern GEJ. A Laboratory Course in Nanoscience and Nanotechnology. CRC Press, Taylor & Francis Group, Boca Raton, FL; 2014

[56] Duan H, Wang D, Li Y. Green chemistry for nanoparticle synthesis. Chemical Society Reviews. 2015;**44**(16):5778-5792

[57] Kumar B, Smita K, Seqqat R, Benalcazar K, Grijalva M, Cumbal L. In vitro evaluation of silver nanoparticles cytotoxicity on hepatic cancer (Hep-G2) cell line and their antioxidant activity: Green approach for fabrication and application. Journal of Photochemistry and Photobiology B: Biology. Jun. 2016;**159**:8-13

[58] Krishnaraj C, Muthukumaran P, Ramachandran R, Balakumaran MD, Kalaichelvan PT. Acalypha indica Linn: Biogenic synthesis of silver and gold nanoparticles and their cytotoxic effects against MDA-MB-231, human breast cancer cells. Biotechnology Reports. 2014;**4**(1):42-49

[59] Dauthal P, Mukhopadhyay M. Phyto-synthesis and structural characterization of catalytically active gold nanoparticles biosynthesized using Delonix regia leaf extract. 3 Biotech. Dec. 2016;**6**(2):118

[60] Ahmed S, Ahmad M, Swami BL, Ikram S. A review on plants extract mediated synthesis of silver nanoparticles for antimicrobial applications: A green expertise. Journal of Advanced Research. Jan. 2016;**7**(1):17-28

[61] Gengan RM, Anand K, Phulukdaree A, Chuturgoon A. A549 lung cell line activity of biosynthesized silver nanoparticles using Albizia adianthifolia leaf. Colloids and Surfaces B: Biointerfaces. May 2013;**105**:87-91

[62] Daisy P, Saipriya K. Biochemical analysis of Cassia fistula aqueous extract and phytochemically synthesized gold nanoparticles as hypoglycemic treatment for diabetes mellitus. International Journal of Nanomedicine. Mar. 2012;**7**:1189-1202

[63] Santhoshkumar T, et al. Synthesis of silver nanoparticles using Nelumbo nucifera leaf extract and its larvicidal activity against malaria and filariasis vectors. Parasitology Research. Mar. 2011;**108**(3):693-702

[64] Narayanan KB, Sakthivel N. Biological synthesis of metal nanoparticles by microbes. Advances in Colloid and Interface Science. Apr. 2010;**156**(1-2):1-13

[65] Klaus T, Joerger R, Olsson E, Granqvist CG. Silver-based crystalline nanoparticles, microbially fabricated. Proceedings of the National Academy of Sciences of the United States of America. Nov. 1999;**96**(24):13611-13614

[66] Beveridge TJ, Murray RG. Sites of metal deposition in the cell wall of Bacillus subtilis. Journal of Bacteriology. Feb. 1980;**141**(2):876-887

[67] Magdi HM, Bhushan B. Extracellular biosynthesis and characterization of gold nanoparticles using the fungus Penicillium chrysogenum. Microsystem Technologies. 2015;**21**(10):2279-2285

[68] Nam KT, et al. Virus-enabled synthesis and assembly of nanowires for lithium ion battery electrodes. Science. May 2006;**312**(5775):885-888

[69] Heath JR. Nanotechnologies for biomedical science and translational medicine. Proceedings of the National Academy of Sciences of the United States of America. Nov. 2015; **112**(47):14436-14443

[70] De Jong WH, Borm PJA. Drug delivery and nanoparticles: Applications and hazards. International Journal of Nanomedicine. 2008;**3**(2):133-149

[71] Rathi Sre PR, Reka M, Poovazhagi R, Arul Kumar M, Murugesan K. Antibacterial and cytotoxic effect of biologically synthesized silver nanoparticles using aqueous root extract of Erythrina indica lam. Spectrochimica Acta Part A: Molecular and Biomolecular Spectroscopy. 2015;**135**:1137-1144

[72] Sharkey RM, Goldenberg DM. Targeted therapy of cancer: New prospects for antibodies and immunoconjugates. CA: A Cancer Journal for Clinicians. 2006;**56**(4):226-243

[73] Wesselinova D. Current major cancer targets for nanoparticle systems. Current Cancer Drug Targets. Feb. 2011;**11**(2):164-183

[74] Liu L, et al. Self-assembled cationic peptide nanoparticles as an efficient antimicrobial agent. Nature Nanotechnology. Jul. 2009;**4**(7):457-463

[75] Maeda H. Macromolecular therapeutics in cancer treatment: The EPR effect and beyond. Journal of Controlled Release. Dec. 2012;**164**(2):138-144

[76] Nichols JW, Bae YH. EPR: Evidence and fallacy. Journal of Controlled Release. Sep. 2014;**190**:451-464

[77] Sharma VK, Siskova KM, Zboril R, Gardea-Torresdey JL. Organic-coated silver nanoparticles in biological and environmental conditions: Fate, stability and toxicity. Advances in Colloid and Interface Science. Feb. 2014;**204**:15-34

[78] Kumar B, et al. One pot phytosynthesis of gold nanoparticles using Genipa americana fruit extract and its biological applications. Materials Science & Engineering. C, Materials for Biological Applications. May 2016;**62**:725-731

[79] Zhou Y, et al. The interactions between ZnO nanoparticles (NPs) and α-linolenic acid (LNA) complexed to BSA did not influence the toxicity of ZnO NPs on HepG2 cells. Nan omaterials (Basel, Switzerland). 2017;**7**(4):1-15

[80] Chang M. Tamoxifen resistance in breast cancer. Biomolecules & Therapeutics. May 2012;**20**(3):256-267

[81] Devulapally R, Sekar TV, Paulmurugan R. Formulation of anti-miR-21 and 4-hydroxy-tamoxifen co-loaded biodegradable polymer nanoparticles and their antiproliferative effect on breast cancer cells. Molecular Pharmaceutics. Jun. 2015;**12**(6):2080-2092

[82] Bouwmeester H, et al. Review of health safety aspects of nanotechnologies in food production. Regulatory Toxicology and Pharmacology. Feb. 2009;**53**(1):52-62

[83] Frohlich E, Roblegg E. Models for oral uptake of nanoparticles in consumer products. Toxicology. Jan. 2012;**291**(1-3):10-17

[84] Wittig A, et al. Amorphous silica particles relevant in food industry influence cellular growth and associated signaling pathways in human gastric carcinoma cells. Nanomaterials (Basel, Switzerland). Jan. 2017;**7**(1):18

[85] Khan S, et al. Evaluation of in vitro cytotoxicity, biocompatibility, and changes in the expression of apoptosis regulatory proteins induced by cerium oxide nanocrystals. Scie nce and Technology of Advanced Materials. 2017;**18**(1):364-373

[86] Gao Y, Chen K, Ma J-L, Gao F. Cerium oxide nanoparticles in cancer. OncoTargets and Therapy. 2014;**7**:835-840

[87] Figueroa CM, Suárez BN, Molina AM, Fernández JC, Torres Z, Griebenow K. Smart release nano-formulation of cytochrome C and hyaluronic acid induces apoptosis in cancer cells. Journal of Nanomedicine & Nanotechnology. Feb. 2017;**1**:8

[88] Akbarzadeh A, et al. Liposome: Classification, preparation, and applications. Nanoscale Research Letters. Feb. 2013;**8**(1):102

[89] Maitz CA, et al. Validation and comparison of the therapeutic efficacy of boron neutron capture therapy mediated by boron-rich liposomes in multiple murine tumor models. Translational Oncology. Jul. 2017;**10**(4):686-692

[90] Sercombe L, Veerati T, Moheimani F, Wu SY, Sood AK, Hua S. Advances and challenges of liposome assisted drug delivery. Frontiers in Pharmacology. Dec. 2015;**6**:286

[91] Sadhu SS, et al. In vitro and in vivo tumor growth inhibition by glutathione disulfide liposomes. Cancer Growth and Metastasis. 2017;**10**:1179064417696070

[92] Drewes CC, et al. Novel therapeutic mechanisms determine the effectiveness of lipid-core nanocapsules on melanoma models. International Journal of Nanomedicine. 2016; **11**:1261-1279

[93] Fior R, et al. Single-cell functional and chemosensitive profiling of combinatorial colorectal therapy in zebrafish xenografts. Proceedings of the National Academy of Sciences. Aug. 2017;**114**(39):E8234-E8243

[94] Mody VV, Siwale R, Singh A, Mody HR. Introduction to metallic nanoparticles. Journal of Pharmacy & Bioallied Sciences. Jun. 2010;**2**(4):282-289

[95] Kumar CSSR. Mixed Metal Nanomaterials. Wiley-VCH Verlag GmbH & Co. KGaA, Weinheim; 2009

[96] Lee JJ, Saiful Yazan L, Che Abdullah CA. A review on current nanomaterials and their drug conjugate for targeted breast cancer treatment. International Journal of Nanomedicine. Mar. 2017;**12**:2373-2384

Chapter 5

Nanotoxicity in Cancer Research: Technical Protocols and Considerations for the Use of 3D Tumour Spheroids

Dania Movia and Adriele Prina-Mello

Additional information is available at the end of the chapter

http://dx.doi.org/10.5772/intechopen.69447

Abstract

The poor clinical translation of oncological nanomedicine products is one of the greatest challenges faced by research today. The use of reductionist *in vitro* models of human cancer and non-predictive animal models is generally considered as one of the main causes of such very low translation rate. The integration of three-dimensional (3D) tumour spheroids in the early stages of the preclinical screening pipeline could significantly facilitate the translation of nanomedicine candidates into clinical practice, by allowing for a more reliable prediction of their efficacy and safety in humans. To lead a successful integration of 3D spheroids, protocols that satisfy issues of ease-of-use, reproducibility and compatibility with conventional and high-throughput assays, without losing the advantages offered by two-dimensional (2D) cell systems, are still needed. To address such need, protocols for the formation and characterisation of scaffold-free 3D tumour spheroids of human adenocarcinoma cells were developed and optimised in this study for their application in nanomedicine safety testing. The protocols reported in this chapter provide the ground on how 3D tumour spheroids could be implemented to design nanomedicine products and speed up experimental cancer research, eliminating those candidates that are likely to be ineffective or unsafe in human at early development stages.

Keywords: 3D tumour spheroids, lung cancer, drug discovery, nanomedicine, safety

1. Introduction

Due to the lack of effective treatment schemes and the high mortality associated with many malignancies, the efforts of the pharmaceutical industry have recently focused on developing new target-oriented nanomedicine products for improving the cancer patients' survival rate [1].

In parallel to novel targeted anticancer drugs (e.g. imatinib, trastuzumab, crizotinib and vemurafenib) [2], various nanomedicine products [3, 4] have been approved for clinical use in recent times, for more effective and safer cancer treatment. Nevertheless, the attrition rate of more than 90% for new drug candidates entering clinical trials is raising increasing concern [5–7]. One of the major reasons for drugs failure, accounting for around 60% of failed clinical trials, is the lack of efficacy in humans [8–10], albeit the therapeutic effect was robustly demonstrated earlier in preclinical studies. The translation rate of nanomedicine products into real clinical anticancer treatments is also worryingly low [3, 11]. Limitations in the preclinical toolbox currently used in the drug discovery pipeline are believed to be one of the major causes for such growing failures and poor translation [5, 12–14]. For example, doubts as to the relevance of animal models in preclinical studies have been raised: various systematic reviews [15–21] describe in fact the inadequateness of animal research for the efficacy assessments of new drug candidates. Conventional two-dimensional (2D) *in vitro* models are likewise considered as highly reductionist [22], endangering the relevance of the preclinical efficacy data collected [5]. Thus, the integration of new preclinical models in the drug development pipeline is urgently needed.

In order to downtrend the raising failure rates in clinical trials, to increase successful clinical translation with reduced R&D costs and animal experimentation, and ultimately to find new and safe cancer therapies (**Figure 1**), preclinical models must better reflect human *in vivo* conditions. This will assist the clear identification of compounds that have the potential to target specifically and selectively those receptors, markers or cellular behaviours characteristic of malignant cells, allowing for newly identified compounds and nanomedicine products to benefit from enhanced efficacy.

There is overwhelming evidence that *in vitro* three-dimensional (3D) tumour spheroids (i.e. microscale 3D spherical cultures of living cancer cells cultured under non-adherent conditions [23] can provide predictive information on drug efficacy and safety in a smart, cost- and time-effective manner [24]. 3D tumour spheroids can be formed by either self-assembling or by forcing cells to grow as cell clusters starting from single cell suspensions [25]. Conventional methods for spheroid formation include hanging drops, culture of cells on non-adherent surfaces, spinner flask cultures and rotary cell culture systems [26].

It has been demonstrated that 3D tumour spheroids more accurately reflect the responses of human tumours than simple 2D cell cultures [27], in particular with respect to drug sensitivity [28, 29] and nanomedicine efficacy [30]. Some targeted compounds have already proven to be more effective in 3D spheroids than in 2D cultures [31, 32]. In general, however, tumour cells cultured in 3D exhibit significantly increased drug resistance compared to those grown in 2D monolayer cultures. For example, in spheroid models of colorectal cancer [33] and pancreatic cancer [34], a reduction in the responsiveness to antitumour

http://dx.doi.org/10.5772/intechopen.69447

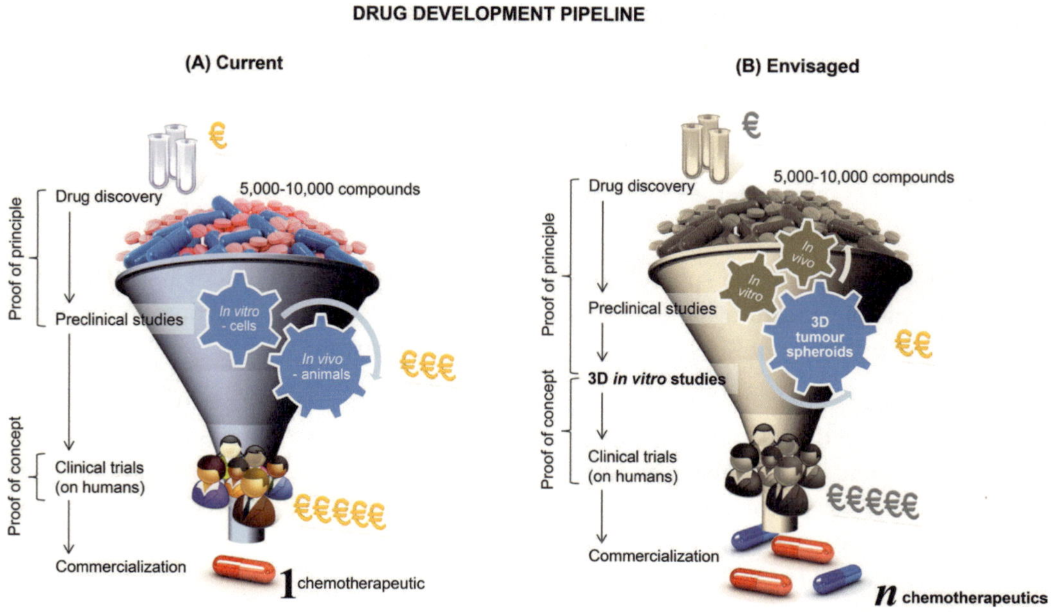

Figure 1. Scheme of the drug development pipeline (A) currently in use and (B) envisaged *via* the implementation of new and more predictive preclinical models. (A) Once active compounds have been identified through the process of drug discovery, preclinical research (*in vitro* studies on cultured cells and *in vivo* studies on animals) are carried out, followed by clinical trials (on humans). Drug development is costly, in particular when carrying out *in vivo* and clinical studies. The current drug development pipeline leads to only one successful market product over 5000–10,000 active compounds identified. (B) Advanced preclinical models capable of predicting drug failures earlier in the 'proof-of-principle' stage, prior to entering costly *in vivo* and clinical studies, will decrease the overall costs of drug development. Studies on 3D tumour spheroids may help significantly, filtering out as early as possible compounds that may not be successful in *in vivo* and clinical studies, thus decreasing the total cost of drug development and increasing the success rate of active compounds in reaching the market as effective chemotherapeutic agents.

agents was observed as a function of multicellular 3D architecture. Multicellular drug resistance is not a new concept and has been studied for the past three decades [35–42]. It is associated with the ability of 3D tumour spheroids from human cancer cell lines to mimic cell dormancy [43, 44]. Cell dormancy originates from the fact that cancer cells in poorly vascularised tumour regions need to adapt to an unfavourable metabolic microenvironment [45], stopping cell cycle progression and becoming dormant. This can confer cancer cells the resistance to drug treatment in humans [46–48]. 3D tumour spheroids can also reproduce cell-cell interactions between cancer cells and/or cancer cells and stroma [49–51]. These interactions strongly influence tumour cells [52] in relation to growth [53], metastasis [53] and response to radio-/chemo-therapy [54]. Various studies have shown in fact that radio- and chemo-resistance of cancer cells is associated with cell-cell adhesion [55]: the term 'cell adhesion-mediated drug resistance' (CAM-DR) has been used to describe this resistance phenomenon [56, 57].

For the purpose of translating oncological nanomedicine products, assessing the ability of the nanocarrier to penetrate into the tumour tissue is essential. 3D tumour spheroids do indeed offer a more predictive *in vitro* model for assessing this aspect [58–65]. Such *in vitro* models have also been extensively used in the last years to evaluate the efficacy [62, 66–72] and safety/toxicity [73–76] of various nanomedicine products with oncological applications, as well as to assess the effect of nanomaterials' physico-chemical properties on their ability to act as nanocarriers of anticancer agents [77–83].

Finally, the use of 3D tumour spheroids considerably limits the ethical issues associated with the use of animal preclinical models, in agreement with the '3Rs (Reduction, Refinement, Replacement) principle' of Russell and Burch in animal research [84].

In light of these considerations, 3D tumour spheroids represent (i) a more predictive and accurate preclinical model and (ii) a key milestone towards a faster and sustainable development of effective nanomedicine products. Consequently, research efforts have recently focused towards adopting 3D tumour spheroids into test platforms, and 3D tumour spheroids has been proposed as 'standard-to-be' for the development and optimisation of new chemotherapeutic agents [85, 86]. From the pharmaceutical industry prospective, it has been also widely accepted that incorporation of 3D tumour spheroids into the drug development pipeline can help selecting the most promising drug candidates prior to clinical trials and to determine future-oriented treatment modalities [14, 87]. What still remains unclear is how to produce such *in vitro* models with a significantly higher degree of equivalence to their *in vivo* counterparts, while making them technically feasible for industrial-scale reliable testing [88]. Indeed, this technological gap has highly hindered the implementation of 3D tumour spheroids as testing platform by the pharmaceutical and nanomedicine industries. Our study aims at addressing this gap by reporting standardised protocols for the formation, characterisation and application of scaffold-free 3D tumour spheroids. Non-small cell lung cancer (NSCLC) was chosen as target cancer type because several therapeutic drugs intended for use in this malignancy have recently failed to show major benefit in clinical trials, despite promising preclinical data [89, 90].

2. Relevance of standardised testing in nanotoxicology

There is overwhelming evidence that *in vitro* 3D tumour spheroids more accurately predict the drug sensitivity of human tumours than conventional 2D cultures. In the market, there are several products available for spheroids preparation; however, for many of them operators are left with the burden to optimise the working protocols to their specific needs. The protocols described herein provide technical solutions for the formation of scaffold-free 3D spheroids and for the characterisation of their architecture and protein marker expression. To support the identification of candidates with clinical potential, protocols are provided for the collection of quantitative data on the efficacy and safety of drug nanocarriers (nanomedicine).

No further optimisation is needed, with the additional advantage of their full compatibility with contemporary high-throughput technologies.

3. Experimental details

3.1. Reagents

Reagents used in this study were chosen due to their wide availability worldwide. Please note that these can be purchased from other distributors and manufacturers than those listed here, except if differently specified. TryplE™ solution, rhodamine phalloidin, Hoechst 33342 counterstain, Mouse Anti-Human β-catenin primary IgG, Alexa Fluor 568-conjugated Goat anti-Mouse secondary IgG, FITC-conjugated Mouse Anti-Human Occludin IgG, Alexa Fluor 488-conjugated Monoclonal Mouse Anti-Human Connexin-43 IgG, Live/Dead Cytotoxicity kit for mammalian cells and 0.4% Trypan Blue Solution (all from Invitrogen Ltd) were purchased from Bio-Sciences Ltd (Ireland). Phosphate-buffered saline (PBS) tablets, 37% paraformaldehyde (PFA) solution, glutaraldehyde (GA), bovine serum albumin (BSA), Triton X-100, hexamethyldisilazane (HMDS) and absolute ethanol (EtOH) were obtained from Sigma-Aldrich (Ireland). VECTASHIELD transparent mounting medium was purchased from Vector Laboratories Inc. (CA, USA). Sheep Anti-Human Fibronectin Antigen Affinity-purified Polyclonal primary IgG, NorthernLights™ 557-conjugated Anti-Sheep secondary IgG, Goat Anti-Human Vimentin Antigen Affinity-purified Polyclonal primary IgG, and NorthernLights™ 557-conjugated Anti-Goat secondary IgG were purchased from R&D Systems (Ireland). FITC-conjugated Mouse Anti-Human E-cadherin IgG and BD Cycletest™ Plus DNA Reagent Kit was supplied by BD Biosciences (Oxford, UK). Connexin 43 antibody and E-cadherin (24E10) rabbit mAb used for Western blot experiments were purchased from Cell Signalling Technology (Brennan and Company, Ireland).

3.2. Cell culture

Human alveolar adenocarcinoma cells (A549 cell line) (ATCC® CCL-185™) and human lung fibroblasts (MRC-5 cell line) (ATCC® CCL-171™) were obtained from the American Tissue Culture Collection (LGC standards, Middlesex, UK). A549 and MRC-5 cells were cultured in Ham's F12K medium supplemented with 2nM L-glutamine (Gibco, Invitrogen Ltd, Bio-Sciences Ltd, Ireland), 1% penicillin/streptomycin (Gibco, Invitrogen Ltd, Bio-Sciences Ltd, Ireland) and 10% foetal bovine serum (FBS) (Sigma-Aldrich, Ireland).

Cell culture flasks, 24-well low-cell binding plates (Nunc™) and 96-well ultra-low attachment (ULA) plates (Corning Costar) were purchased from Fisher Scientific (Ireland). Happy Cell™ ASM medium and 96-well low-cell binding plates (Biocroi Ltd) were kindly donated by Biocroi Ltd (Ireland). Happy Cell™ ASM medium is a polymer-based suspension media of low viscosity that enables the 3D culture of cells [91]. 96-well flat-bottom, non-treated plates (BD Falcon™) and 96-well U-bottom, non-treated plates (BD Falcon™) were purchased from BD Biosciences (Oxford, UK). Four-well Millicell EZ slides were supplied by Millipore (Ireland).

3.3. Equipment

A Zeiss 510 meta laser scanning confocal microscope (LSCM) equipped with a Zeiss LSM 5 software and Zeiss Orion Plus He-ion microscope (both from Carl Zeiss, Germany) were used for imaging the 3D tumour spheroids. A Countess™ cell counter (Invitrogen, UK) was used for trypan blue exclusion assay, while BD Accuri C6 flow cytometer (Becton Dickinson Biosciences, UK) was used for high-throughput assays. The Volocity 3D Image Analysis Software (PerkinElmer Inc., MA, USA) was used for surface rendering of Z-stack images and co-localization studies. Flow cytometry was carried out by means of BD Accuri™ C6 flow cytometer (BD Biosciences, Oxford, UK).

4. Technical protocols and considerations

This study describes the protocols for the formation, characterisation and use of 3D tumour spheroids as *in vitro* models for testing nanomedicine products. These protocols, which are simple, validated and reproducible, can be grouped into three main categories (highlighted in blue boxes in **Figure 2**): (1) protocols for the formation of 3D tumour spheroids, (2) protocols for analytical techniques and (3) protocols for specialised applications.

4.1. Formation of 3D tumour spheroids

One aspect commonly missing in all commercially available products for the formation of scaffold-free 3D tumour spheroids is the description of achievable spheroid size. In addition to this, not all available methodologies produce an abundance of 3D spheroids, and often the *in vitro* models formed have different dimension and shape. The direct consequence is that 3D tumour spheroids are often not comparable among studies, as they are formed of cells in different proliferative and metabolic states, raising serious concerns about the reproducibility of data produced. A recent study has in fact showed that a number of morphology parameters (including volume and shape) affect the response of spheroids to treatment [92]. Such lack of reproducibility is hindering the use of 3D tumour spheroids in preclinical tests.

A protocol was developed in this study allowing for the formation of scaffold-free, non-adherent 3D tumour spheroids of A549 cells with or without the use of a reference commercial product (Happy Cell™ ASM) in combination with various commercially available multi-well plates. Our protocol allowed forming 3D models with well-defined and highly reproducible size in a range between 200 μm and 2 mm. In particular, our 3D tumour spheroids mimicked the size of early stage NSCLC at clinical stages 0–I, where the tumour mass has dimensions below 3 cm.

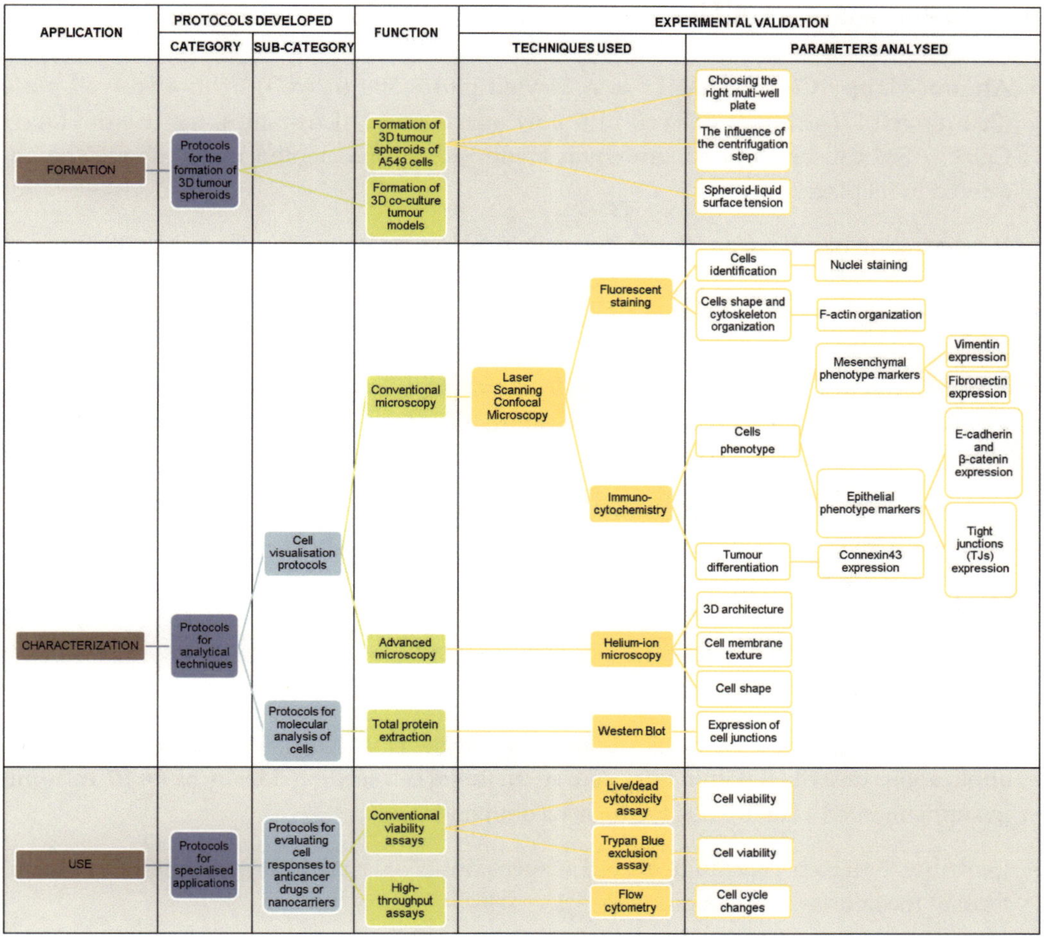

Figure 2. Scheme composing the main application of the protocols developed in this study. Protocols are grouped into three main categories. In some cases, sub-categories can be identified. The protocols developed were validated for their function through experiments analyzing various cellular parameters, by means of standard well-established techniques.

4.1.1. Formation of 3D tumour models of A549 cells

3D tumour spheroids of A549 cells were prepared as described here below.

A. Preparing Happy Cell™ ASM for use:

- Dilute F12K/DMEM Happy Cell™ ASM 2× (commercial product) in Ham's F12K media (supplemented with 2nM L-glutamine, 1% penicillin-streptomycin and 10% foetal bovine serum (FBS)) in 1:9 ratio, thus obtaining Happy Cell™ ASM 1×.

B. Preparing multi-well plates for cell seeding:

- Aliquot Happy Cell™ ASM 1× in a 24-well plate (500 μl/well) or in a 96-well plate (200 μl/well). If using multi-well ultra-low attachment (ULA) plates, the use of Happy Cell™ ASM can be avoided. An equal amount of supplemented media should be aliquoted in its place.
- If preparation of cell suspension is delayed, incubate plates at 37°C, 5% CO_2 until use.

C. Cell seeding:

- Maintain cell line as adherent monolayer cultures in T75 cell culture flasks in Ham's F12K media supplemented with 2 nM L-glutamine, 1% penicillin-streptomycin and 10% FBS. Incubate at 5% CO_2 at 37°C.
- Rinse the cells with phosphate buffered saline (PBS) and add 2 mL of TrypLE™ to detach adherent A549 cells (at 80% confluence) from the culture flask.
- Incubate at 37°C for 3–5 min.
- Neutralise TrypLE™ with 6 mL fresh supplemented Ham's F12K media and centrifuge the cells suspension at 5000 rpm for 4 min.
- Aliquot the cell suspension in 1.5 mL Eppendorf tubes (1 mL cells suspension/tube). For the applications described within this exercise, an initial cell seeding density of 1×10^6 cells/mL in supplemented Ham's F12K media is recommended.
- Centrifugation step: centrifuge the cell suspensions (7200 rpm, 1 min). A cells pellet should form at the bottom of the microtube.
- Carefully aspirate one cells pellet at a time with a sterile, plastic 1000 μL syringe tip.
- Inoculate the cells pellet on the Happy Cell™ ASM 1× (or supplemented media) previously aliquoted in multi-well plates (1 cells pellet/well).
- Incubate for 4 days at 37°C and 5% CO_2. This culture time (spheroid growth phase) is comparable to those reported in the scientific literature [93].

Our experimental data indicated that three main parameters are key in defining the final size of the 3D tumour spheroids formed following our protocol. These are (i) competitive cell adhesion and cell monolayer formation (the stronger is the cell adhesion to the culture substrate, the smaller are the 3D tumour spheroids formed); (ii) initial cell aggregation (the higher cell aggregation is when seeding the cells, the bigger are the 3D tumour spheroids formed) and (iii) the spheroid-liquid surface tension (the lower is the viscosity of the spheroid's surrounding environment, the bigger the spheroids will grow in size). In the next sections, detailed evidence on the importance of such parameters is presented and supported by experimental data.

http://dx.doi.org/10.5772/intechopen.69447

4.1.1.1. Competitive cell adhesion and cell monolayer formation: selecting the appropriate multi-well plate

Commercially available multi-well plates are available in a variety of formats. In deciding which multi-well plate format to use, the application intended for 3D tumour spheroids should be considered carefully.

The overall tumour response to chemotherapeutic treatments is in fact affected by a multitude of factors, among which is the tumour size. Tumour size is known to strongly affect the diffusion and penetration of molecular and nano-enabled chemotherapeutic treatments decreasing their efficacy [94]. By varying their size, 3D tumour spheroids are thus amenable to therapy-related studies with different emphasis, ranging from studies focusing either on the investigation of the micro-environmental regulation of tumour cell physiology or on the therapeutic efficacy of drugs in authentic pathophysiological milieu conditions. Spheroid size needs therefore to be precisely defined when integrating 3D cell models in drug-testing strategies.

Several commercially available multi-well cell culture plates compatible with high-throughput assays have been tested within this study. In order to perform a comparative assessment of the most suitable multi-well plates for spheroid formation, identical spheroid preparation reagents were used in all relevant experiments. Thus, 3D tumour spheroids were always formed in Happy Cell™ ASM, in combination with various commercially available multi-well plates. The growth of 3D tumour spheroids was monitored by light microscopy for all the tested multi-well plate formats. **Table 1** reports the size distributions of the spheroids obtained in the various plate formats investigated, whereas representative light microscopy images of the 3D tumour spheroids formed are shown in **Figure 3**. Light microscopy imaging showed that 3D tumour spheroids formed in 24-well low-cell binding plates and ULA plates were bigger and more compact, and therefore, more mechanically robust.

4.1.1.2. Initial cell aggregation: the influence of centrifugation on the size of 3D tumour spheroids

Next important feature to consider is how cell aggregation during seeding influences the final size of the 3D tumour spheroids formed. A parallel set of experiments, where cells were seeded avoiding any centrifugation prior to plating, demonstrated the strong influence of this step in defining the final size of the spheroids, except when using 96-well low-cell binding plate. The main outcomes of our experiments (in terms of tumour size) are summarised in **Table 2**. Our results demonstrated that, for obtaining spheroids of size above 1 mm after 4 days in culture, the centrifugation step described in the protocol was necessary.

4.1.1.3. Spheroid-liquid surface tension

As stated above, the viscosity of the spheroid's surrounding environment influences the final size of the *in vitro* model. Our experimental data showed that the dilution factor of Happy Cell™ ASM in supplemented cell culture media contributed to defining the final size of the 3D tumour spheroids formed (**Table 3**). In detail, increasing the viscosity of the surrounding

Multi-well plate format				Spheroid size (mm)	Notes
Number of wells	**Well shape**	**Well surface material**	**Supplier**		
4	Flat-bottom	Glass	Millipore	Not formed	• High cell adhesion • Not suitable
24	Flat-bottom	Low-cell binding polystyrene	Nunc™	1.5 ± 0.5	• Optimal substrate • No cell adhesion • Mechanically robust spheroids
96	Flat-bottom	Low-cell binding polystyrene	Biocroi Ltd	0.18 ± 0.11	• Partial cell adhesion
		ULA polystyrene	Corning Costar	0.85 ± 0.55	• Optimal substrate • No cell adhesion • Mechanically robust spheroids • 3D spheroids size without Happy Cell™ ASM: 0.3–1.3 mm
		Non-treated polystyrene	BD Falcon™	0.4 ± 0.1	• Partial cell adhesion
	U-bottom	Non-treated polystyrene	BD Falcon™	0.35 ± 0.25	

Tumour spheroids were grown in Happy Cell™ ASM at a 1:9 (Happy Cell™ ASM: supplemented media) dilution ratio. With exemption of 96-well plates from Biocroi Ltd., all multi-well plate formats tested are gold-standard cell culture substrates available from conventional suppliers. Cell culture plates provided by Biocroi Ltd. are included for comparison, as a recommended substrate for spheroids growth in Happy Cell™ ASM.

Table 1. Multi-well plate formats tested and main experimental outcomes in respect to the size of 3D tumour spheroids formed.

environment by reducing the dilution of Happy Cell™ ASM to a 1:1 (Happy Cell™ ASM: supplemented media) ratio caused a reduction in tumour spheroids size, except in ULA plates. Thus, our results suggested that, for obtaining spheroids of size above 1 mm, a 1:9 (Happy Cell™ ASM: supplemented media) ratio or the use of ULA multi-well plates (in combination or without Happy Cell™ ASM) is recommended.

4.1.2. Formation of 3D co-culture tumour models

3D co-culture tumour models including heterotypic cellular components (also referred to as multicellular spheroid (MCS) models or multicellular 3D tumour spheroids) play a critical role in recreating the tumour microenvironment *in vitro*. The tumour microenvironment plays a critical role in cancer cell differentiation, and greatly impacts therapeutic efficiency of chemotherapeutic drugs. Co-culture 3D tumour models represent therefore one of the most promising *in vitro* systems for predictive testing of compound efficacy in oncology [94]. Thus, in this study, we developed protocols allowing the formation of a 3D co-culture tumour model that aims to analyse the interplay of NSCLC cells and the healthy surrounding connective tissue. In detail, our model comprised 3D tumour spheroids of lung cancer epithelial

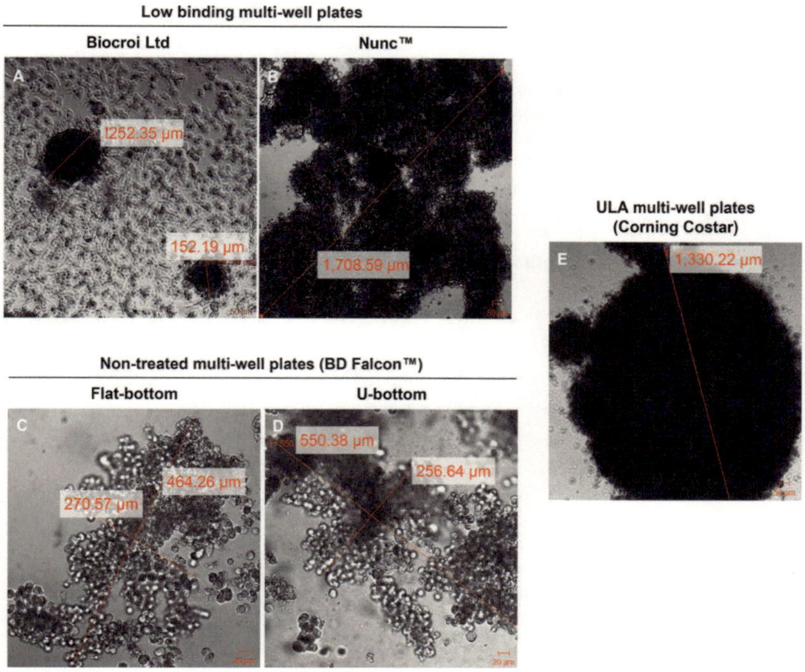

Figure 3. Comparison of 3D tumour spheroids size formed in various multi-well plate formats: (A) 96-well low-cell binding plate; (B) 24-well low-cell binding plate; (C–D) 96-well non-treated plate with (C) flat-bottom or (D) U-bottom wells; and (E) 96-well ULA plate. Tumour spheroids were grown in Happy Cell™ ASM at a 1:9 (Happy Cell™ ASM: supplemented media) dilution ratio.

Multi-well plate format				Spheroid size (mm)	Consequences of lack of centrifugation step
Number of wells	**Well shape**	**Well surface material**	**Supplier**		
4	Flat-bottom	Glass	Millipore	Not formed	• None
24	Flat-bottom	Low-cell binding polystyrene	Nunc™	0.75 ± 0.5	• Decreased spheroids size
96	Flat-bottom	Low-cell binding polystyrene	Biocroi Ltd	0.14 ± 0.06	• No significant changes in spheroids size
		ULA polystyrene	Corning Costar	0.23 ± 0.16	• Decreased spheroids size
		Non-treated polystyrene	BD Falcon™	0.15 ± 0.05	• Decreased spheroids size
	U-bottom	Non-treated polystyrene		0.08 ± 0.02	

Spheroids were grown in Happy Cell™ ASM at a 1:9 (Happy Cell™ ASM: supplemented media) dilution ratio. Conclusions on the consequences associated to the lack of the centrifugation step on the spheroids size are drawn based on the comparison to the spheroid sizes values reported in **Table 1**.

Table 2. Influence of centrifugation step on the final size of the 3D tumour spheroids formed in the various multi-well culture plates tested.

Multi-well plate format				Spheroid size (mm)	Influence of increased viscosity of environment
Number of wells	Well shape	Well surface material	Supplier		
4	Flat-bottom	Glass	Millipore	Not formed	• None
24	Flat-bottom	Low-cell binding polystyrene	Nunc™	1.05 ± 0.15	• Decreased spheroids size
96	Flat-bottom	Low-cell binding polystyrene	Biocroi Ltd	0.15 ± 0.05	• No significant changes in spheroids size
		ULA polystyrene	Corning Costar	1.25 ± 0.75	• Decreased spheroids size
		Non-treated polystyrene	BD Falcon™	0.15 ± 0.05	• Decreased spheroids size
	U-bottom	Non-treated polystyrene		0.2 ± 0.1	

The experimental outcomes (in terms of tumour size) are here reported for spheroids grown in Happy Cell™ ASM at a 1:1 (Happy Cell™ ASM: supplemented media) dilution ratio. The influence of the increased viscosity of the environment on the spheroids size are evaluated based on the comparison to the spheroid sizes values reported in **Table 1**.

Table 3. Influence of the viscosity of the surrounding environment on the final size of the 3D tumour spheroids formed.

(A549) cells cultured on a 2D monolayer of healthy stromal fibroblasts (MRC-5 cells). 3D co-culture tumour models were prepared as described below.

A. Formation of a fibroblast monolayer:

- Maintain MRC-5 cell line as adherent monolayer culture in T75 cell culture flasks in Eagle's Minimum Essential Medium (EMEM) media supplemented with 1% penicillin-streptomycin and 10% FBS, at 5% CO_2 at 37°C.
- Rinse the cells with PBS and add 2 mL of TrypLE™ to detach adherent MRC-5 cells (at 80% confluence) from the culture flask.
- Incubate at 37°C for 3–5 min.
- Neutralise TrypLE™ with 6 mL fresh supplemented EMEM media and centrifuge the cells suspension at 5000 rpm for 4 min.
- Using cell seeding density of 1×10^6 cells/mL in supplemented EMEM media, plate the cells in four-well chambered Millicell EZ slides (final volume: 500 μL/well).
- Incubate for 2/3 days at 37°C and 5% CO_2, until cell monolayer confluence is achieved.

B. Seeding of 3D tumour spheroids:

- With a sterile, plastic 1000 μL syringe tip, carefully aspirate the 3D tumours spheroids previously grown from A549 cells for 4 days with or without Happy Cell™ ASM 1×.

- Inoculate the 3D tumour spheroids on the fibroblast monolayers. It is recommend to use a mixture of supplemented Ham's F12K and EMEM media (ratio 1:1) as culture media for the co-culture 3D tumour models.
- Incubate for 24 h at 37°C and 5% CO_2 to allow the formation of cell-cell adhesions.

Figure 4 shows representative laser scanning confocal microscopy (LSCM) images of the fully formed 3D co-culture tumour models. For specific LSCM staining and imaging protocols please refer to the appropriate sections of this manuscript.

4.2. Protocol for analytical techniques

A wide variety of techniques can be used to study 3D tumour spheroids characteristics. A comprehensive description of such techniques can be found in recently published reviews [26, 95]. The protocols that we describe within this section provide practical guidelines for the application of some of these techniques, allowing for the analytical characterisation of 3D tumour spheroids and 3D co-culture tumour models. These protocols were optimised on 3D tumour spheroids grown in 96-well ULA plates (with or without Happy Cell™ ASM), as these proven to be the most mechanically robust cultures formed. Robustness of 3D tumour spheroids was assessed as for their capability to maintain their shape and size under mechanical stress (e.g. agitation or transfer with a pipette). Conventional two-dimensional (2D) cell monolayers (grown onto glass substrates according to cell supplier guidelines) were used as controls during the validation of our protocols for scaffold-free 3D spheroids.

4.2.1. Cellular imaging

A complete characterisation of the spheroid architecture and protein markers expression in the 3D tumour models used in preclinical studies is crucial for extrapolating and interpreting

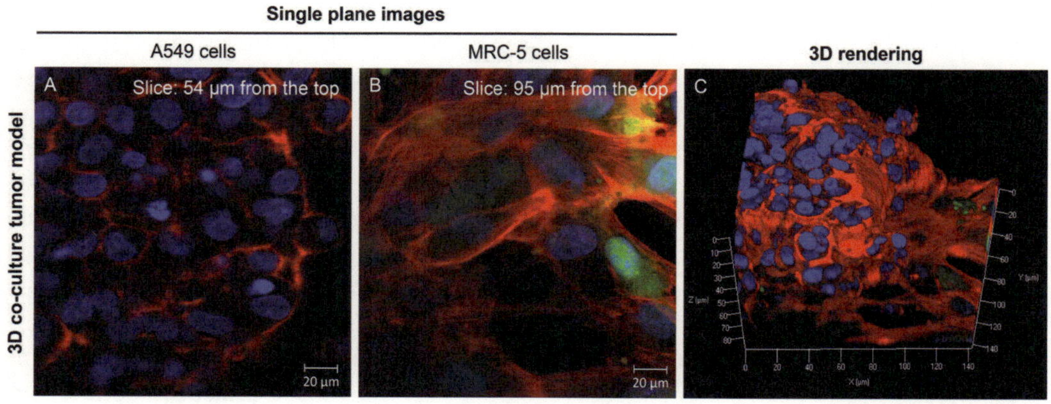

Figure 4. Representative LSCM images of a 3D co-culture tumour model stained for F-actin (rhodamine phalloidin; 1:40 dilution; in red) and nuclei (Hoechst; 1:400 dilution; in blue). MRC-5 cells were also stained with 20 µM Cell Tracker Green CMFDA (in green) for 45 min (37°C and 5% CO_2) prior to seeding the 3D spheroids, to allow their easy identification in the co-culture model. Images were collected in (A–B) single-plane or (C) Z-stack mode with an oil-immersion 63× objective lens. (C) 3D rendering of Z-stack images obtained with the 3D function of Zeiss ZEN software (Carl Zeiss, Germany) (114 sections, total height: 102 µm). (A–C) Scale bars: 10 µm.

valuable data on the efficacy of the molecular or nano-enabled chemotherapeutic agent under investigation. Because scaffold-free 3D tumour spheroids are grown in liquid media and do not adhere to a substrate, such cell models are difficult to image due to the Brownian motion in the culture plates. Thus, the first challenge in imaging scaffold-free tumour spheroids is to stabilise the 3D cell structures, without disrupting their architecture and markers expression, to allow high-quality images to be captured and analysed.

In this study, protocols are described for monitoring and analysing fixed, physically intact spheroid cultures. By applying these protocols it is possible to implement a suite of conventional and advanced imaging technologies, such as LSCM and He-ion microscopy (HIM) to characterise: (i) cells' shape and organisation of their cytoskeleton; (ii) phenotype of the cells forming the 3D tumour spheroids and (iii) the degree of tumour differentiation.

Essential steps needed for preparing 3D tumour spheroids for conventional or advanced imaging analysis are described below.

A. Harvesting:

- At the chosen time point, harvest cells by gentle aspiration of 100 μL/well media. This operation cannot be performed with vacuum pipettes, to avoid accidental aspiration of the 3D tumour spheroids formed.

B. Fixation:

- Add 200 μL/well of fixative and incubate at room temperature.
- Special considerations: fixation of 3D tumour spheroids can be achieved by a chemical approach. Fixation is usually the first stage in a multistep process to prepare a sample for microscopy or other analysis. It is important, when selecting a fixative to have clearly set in mind the final purpose of the analysis, which must be considered and matched by the requirements for the analytical technique. Some fixatives are suitable for general structure analysis, others for immunocytochemistry. Survival of tissue antigens for immunochemical staining depends on the type and concentration of fixative, on fixation time and on the size of the tissue specimen to be fixed. In relation to the suite of conventional and advanced imaging technologies discussed in this study, it is recommended the use of 4% paraformaldehyde (PFA) as fixative for conventional imaging techniques (e.g. LSCM), with an incubation time of 10 min at room temperature. If immunocytochemistry is performed, the selection of the fixative solution should be in accordance with the manufacturer's instructions for the antibodies used. Finally, for advanced imaging analysis (e.g. HIM), immersion in 2.5% glutaraldehyde (GA) for 30 min at room temperature (25°C) is recommended. Incubation overnight at 4°C is also suitable. Avoid longer incubation time, as formaldehyde-derived products can cause cells shrinkage overtime.
- Gently aspirate 200 μL/well.

C. Washing:

- Wash with 200 μL/well PBS by gently adding and then removing the solution to/from the wells.
- Agitation or shacking is to be avoided during this phase to not shear the 3D spheroids from the bottom of the well.

D. Mounting:

- Gently aspirate the 3D tumour spheroids with a 1000 μL tip and transfer them on microscope glass slides.
- Mount the glass slide in transparent mounting medium prior to LSCM analysis—when covering with a glass coverslip, take care not to create air bubbles, then seal with nail varnish or tape. If using nail varnish, leave to dry for at least 30 min at room temperature in the dark before imaging.

E. Storage:

- The stability of the 3D spheroids has been tested at regular intervals: fixed spheroids can be stored in sterile PBS at 4°C up to 1 month.
- Glass slides can be stored in the dark at 4°C up to few months.

4.2.1.1. Conventional microscopy

When viewing an unstained 3D tumour spheroid under brightfield illumination, the combined density and thickness of the 3D culture prevents the clear visualisation of individual cells (as seen in **Figure 3E**). However, fluorescent staining and immunocytochemistry coupled with LSCM can be successfully used to image 3D tumour spheroids at the individual cell level. LSCM relies in fact on the combination of point illumination and a pinhole to eliminate most of the out-of-focus light signal and allows for reconstruction of 3D volumes, making it ideal to image thick samples, such as 3D tumour spheroids.

Here we describe the protocols for the visualisation of various cell markers within 3D tumour spheroids by fluorescent staining or immunocytochemistry. For best results we recommend to acquire Z-stack images by LSCM with intervals in the range between 0.8 and 1.2 μm. Note that spheroids thickness might in some cases exceed the Z-stack capacity of some models of confocal microscopes. Please also note that successful LSCM imaging requires the careful optimization of microscope set-ups. Specific filters, detectors and pinhole size for imaging the specimens might need to be optimised according to the specifications of the microscope and fluorescent dyes used. Long-term exposure of dyes to fluorescent light can lead to photobleaching, so ensuring that shutters closed in between image acquisitions can reduce this problem.

3D tumour spheroids of A549 cells can be stained with fluorescent probes or by immunocytochemistry as described below.

A. Fluorescent staining of fixed 3D tumour spheroids:

- Add staining solution (100 μL/well) to the specimens.
- Incubated for 4 h at room temperature in the dark.
- Wash with 200 μL/well PBS by gently adding and then removing the solution to/from the wells. Repeat the washing step twice.

B. Immunocytochemistry of fixed 3D tumour spheroids

- If a cell permeabilization step is needed, incubate overnight at 4°C with 0.1% Triton X-100 (200 μL/well). The long incubation time is needed to allow the solution to perfuse into the inner core of the 3D tumour spheroids. Note that if using methanol/acetone fixation, no permeabilization step is required.
- Wash with 200 μL/well PBS by gently adding, and then removing the solution to/from the wells.
- Add 200 μL/well blocking buffer (1% BSA) and incubate overnight at 4°C, for avoiding the antibody unspecific binding. It is recommended that for each antibody used, the original manufacturers' instructions regarding blocking reagents must be followed.
- Remove the blocking buffer by gently aspirating 200 μL/well and wash with PBS as previously described (washing step).
- Add the primary antibody (100 μL/well) previously prepared in 0.1% Triton X-100 solution.
- Incubate for 4 h at room temperature. If the antibody is already conjugated to a fluorophore, protect from light.
- Repeat the washing step twice.
- If an unconjugated primary antibody was used, add the secondary (labelled) antibody (100 μL/well) and incubate for 4 h at room temperature in the dark.
- Repeat the washing step twice.

4.2.1.1.1. Experimental validation: cell shape and cell cytoskeleton organisation—F-actin staining

Our protocol for fluorescent staining proved to be useful in quantifying the effect of the surrounding environment when culturing cells in 3D. Similarly to signalling molecules, the mechanical stimuli applied by the surrounding environment to cells induce subcellular and cellular events, such as cytoskeleton remodelling and cell shape changes [96, 97]. 3D tumour spheroids were formed without Happy Cell™ ASM in 96-well ULA plates, stained for F-actin (which is one of the main components of cells' cytoskeleton) and analysed by LSCM. Representative LSCM images of 3D tumour spheroids are shown in **Figure 5**. A549

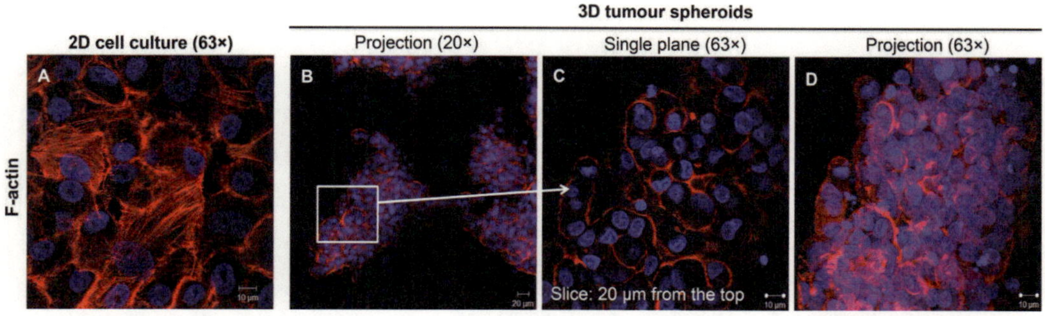

Figure 5. LSCM images of A549 cells cultured in (A) 2D and (B–D) 3D. Cells were stained with rhodamine phalloidin (1:40 dilution) to detect F-actin filaments (in red) and Hoechst (1:400 dilution) as nuclear counterstain (in blue). Images were collected in (A, C) single-plane or (B, D) Z-stack mode with a 20× (scale bar: 20 μm) or an oil-immersion 63× (scale bars: 10 μm) objective lens. (D) Projection of Z-stack images obtained with the projection function of Zeiss LSM 5 software (61 sections, total height: 48 μm).

cells grown in 2D on glass substrates and stained for F-actin were used as control. Our protocol allowed determining that, when grown in scaffold-free 3D tumour spheroids, A549 cells showed a round cell shape and expressed cortical F-actin; in contrast, 2D cell cultures showed a cytoskeleton organised in cytoplasmic F-actin stress fibres.

4.2.1.1.2. Experimental validation: cell phenotype—expression of mesenchymal and epithelial markers

The microenvironment is known to influence the conversion of epithelial cells into mesenchymal cells in *in vitro* systems [98, 99] by a process called epithelial-to-mesenchymal transition (EMT). EMT is a critical series of events that switch early stage carcinomas into invasive malignancies. EMT is associated with the loss of epithelial cell markers and the acquisition of mesenchymal features [100, 101].

By applying our protocol for immunocytochemistry, 3D tumour spheroids formed in 96-well ULA plates were assessed for the expression of two mesenchymal markers: vimentin and fibronectin. **Figure 6** shows some representative results. In clinical trials, vimentin expression is used as a clinical marker of the response of NSCLC to chemotherapeutic agents [102], since increased expression of this protein gives an indication of tumour progression. In parallel, fibronectin expression is increased in NSCLC, enhancing the cells' invasiveness and conferring them resistance to apoptosis-inducing chemotherapeutic agents. Thus, the detection of these markers was selected during validation experiments as it finds a useful application in the preclinical drug efficacy-screening pipeline. Our immunocytochemistry could, for example, be used for characterising spheroids cultured at different time points or formed by different protocols, allowing for the selection of the most relevant *in vitro* model to be used in preclinical tests. The expression levels to be mimicked should be assessed based on patient-based, clinically relevant data.

Our immunocytochemistry protocol was also found to be applicable for the evaluation of the expression of cleaved E-cadherin and β-catenin and for the investigation of their localization in the cell body (**Figure 7A–F**). Cells with epithelial phenotypes form adherent junctions. In such junctions, β-catenin binds to the cytoplasmic tail of the protein E-cadherin. During EMT,

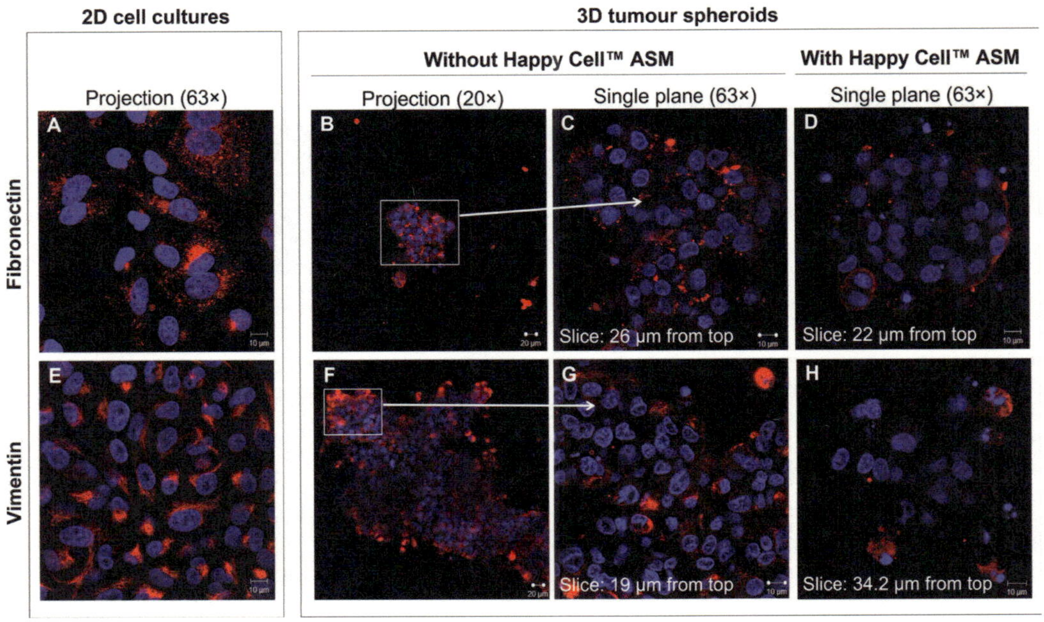

Figure 6. LSCM images of (A, E) 2D cell cultures and (B–D, F–H) 3D tumour spheroids stained for (A–D) fibronectin and (E–H) vimentin (both in red). 3D tumour spheroids were formed in ULA plates (B, C, F, G) with or (D, H) without Happy Cell™ ASM. For fibronectin detection, cells were stained with Sheep Anti-Human Fibronectin Antigen Affinity-purified Polyclonal primary IgG (1:20 dilution) and NorthernLights™ 557-conjugated Anti-Sheep secondary IgG (1:200 dilution) was used as secondary antibody. For vimentin immunocytochemistry, cells were stained with Goat Anti-Human Vimentin Antigen Affinity-purified Polyclonal primary IgG (1:20 dilution) and NorthernLights™ 557-conjugated Anti-Goat secondary IgG (1:200 dilution) was used as secondary antibody. Nuclei were counterstained with Hoechst (1:400 dilution) (in blue). (A, E) In 2D cell cultures, fibronectin showed a dotted pattern, while vimentin appeared as robust stress fibres. (B–D, F–H) Cells in 3D tumour spheroids showed cytoplasmic expression of vimentin and fibronectin, with reduced expression in centrally located cells (spheroid core). (B, F) Projection of Z-stack images obtained with the projection function of Zeiss LSM5 software (19 sections, total height: 46.77 μm). (B, F) White boxes highlight the region of the 3D tumour spheroids magnified in image (C) and (G), respectively. Scale bars: (B, F) 20 μm (20× objective lens) and (A, C–E, G, H) 10 μm (63× objective lens).

the intracellular fragment of E-cadherin (called cleaved E-cadherin) translocates into the cell nucleus. This results in the abolishment of the E-cadherin-mediated sequestering of β-catenin in the cytoplasm and the translocation of β-catenin to the cell nucleus. Co-localization studies were possible on an example set of LSCM images acquired from specimens prepared as described above. Such studies showed that cleaved E-cadherin and β-catenin were mainly localised in the cells' nuclei when A549 cells were cultured as 3D tumour spheroids in 96-well ULA plates without Happy Cell™ ASM (**Figure 7J**).

Tight junctions (TJs) are also involved in cell adhesion and lung cancer development [103]. In particular, occludin plays a critical role in defining the cellular phenotypes in solid tumours [104]: while epithelial cell phenotypes express occludin, cells with mesenchymal phenotypes downregulate occludin expression, resulting in enhanced cellular invasiveness and motility and thus promoting tumorigenic and metastatic properties of tumour cells [101]. **Figure 8** shows representative LSCM images demonstrating that the immunocytochemistry protocol described in this study allowed evaluating the expression of occluding in 3D tumour spheroids.

http://dx.doi.org/10.5772/intechopen.69447

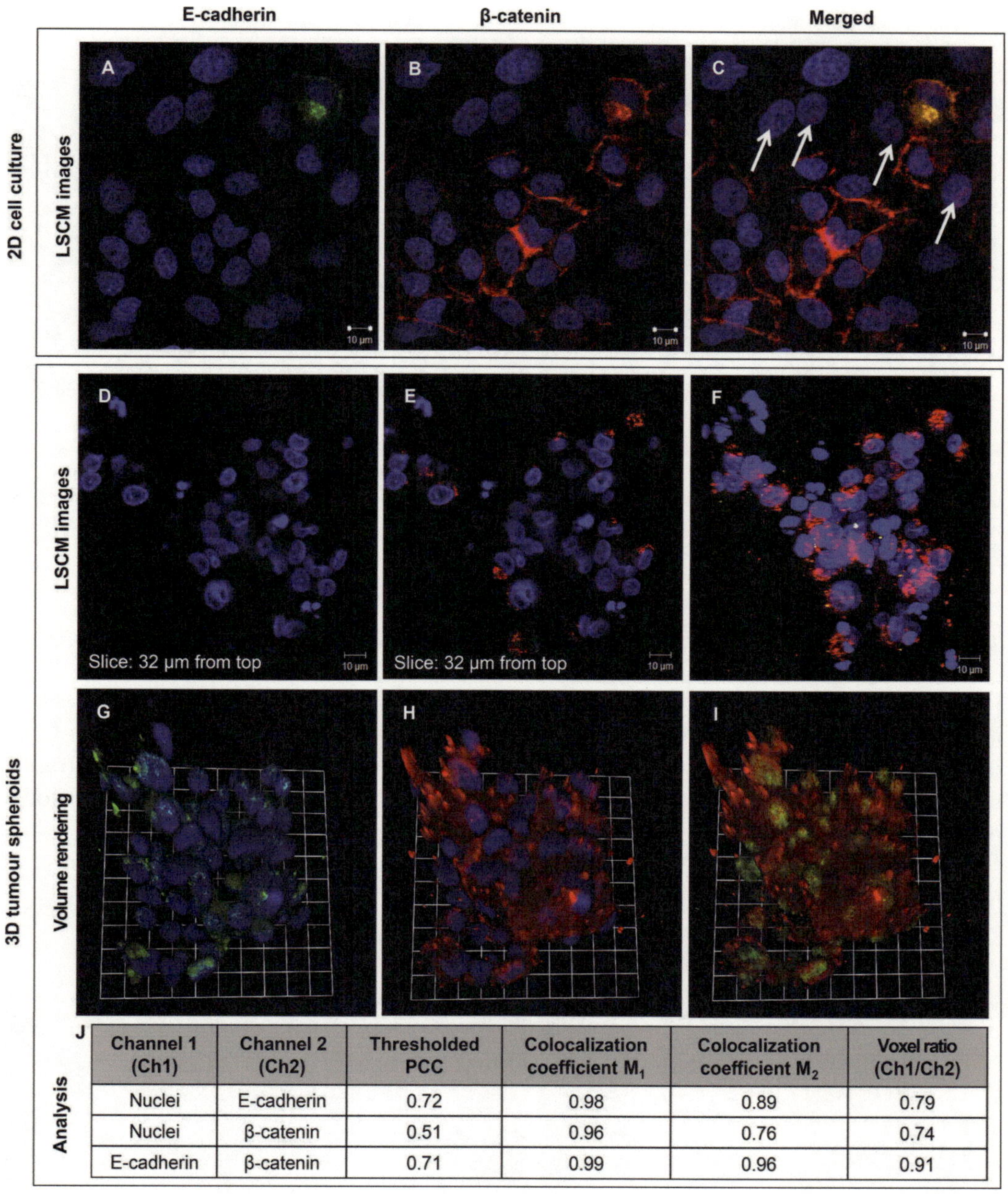

J

Analysis

Channel 1 (Ch1)	Channel 2 (Ch2)	Thresholded PCC	Colocalization coefficient M_1	Colocalization coefficient M_2	Voxel ratio (Ch1/Ch2)
Nuclei	E-cadherin	0.72	0.98	0.89	0.79
Nuclei	β-catenin	0.51	0.96	0.76	0.74
E-cadherin	β-catenin	0.71	0.99	0.96	0.91

Figure 7. LSCM images of (A-C) 2D cell cultures and (D–F) 3D tumour spheroids stained for cleaved E-cadherin (in green), β-catenin (in red) and nuclei (in blue). 3D tumour spheroids were formed in ULA plates without Happy Cell™ ASM. A549 cells were stained with FITC-conjugated Mouse Anti-Human E-cadherin IgG (1:70 dilution), Mouse Anti-Human β-catenin primary IgG (1:50 dilution) and Hoechst nuclear counterstain (1:400 dilution). As secondary antibody, Alexa Fluor 568-conjugated Goat anti-Mouse secondary IgG (1:1000 dilution) was used. (C) White arrows highlight nuclear translocation of β-catenin. (C, F) Projection of Z-stack images obtained with the projection function of Zeiss LSM 5 software (54 sections, total height: 57.32 μm). Scale bars: 10 μm (63× objective lens). (G–I) Volume rendering of representative Z-stack LSCM images of 3D tumour spheroids and (J) co-localization studies. (G–I) Volume rendering was obtained with the Volocity software. Scale bar: 1 unit = 12 μm.

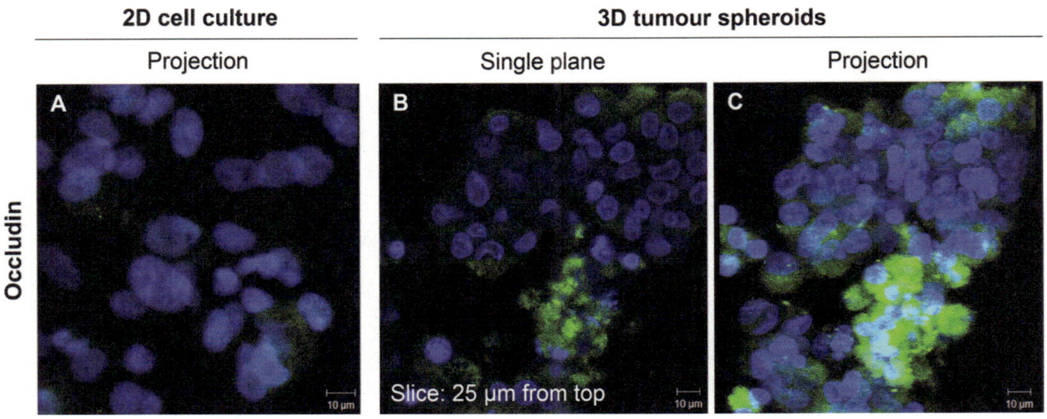

Figure 8. LSCM images of (A) 2D cell cultures and (B–C) 3D tumour spheroids stained for occludin with FITC-conjugated Mouse Anti-Human Occludin IgG (1:50 dilution) (in green) and for nuclei with Hoechst (1:400 dilution) (in blue). 3D tumour spheroids were formed in ULA plates, without Happy Cell™ ASM. (A, C) Projection of Z-stack images were obtained with the projection function of Zeiss LSM 5 software (77 sections, total height: 76 μm). (A–C) Scale bars: 10 μm (63× objective lens).

4.2.1.1.3. Experimental validation: tumour differentiation—expression of Connexin-43 (Cx-43)

Cx43 represents one of the predominant gap junction proteins; its down-regulation is associated to poorly differentiated NSCLC, as showed *in vitro* and in human tissue [105]. Thus, Cx43 is a cellular marker of the tumour differentiation. **Figure 8** shows representative micrographs of fixed 3D tumour spheroids labelled with an anti-Cx43 antibody following our protocol described above (**Figure 9**).

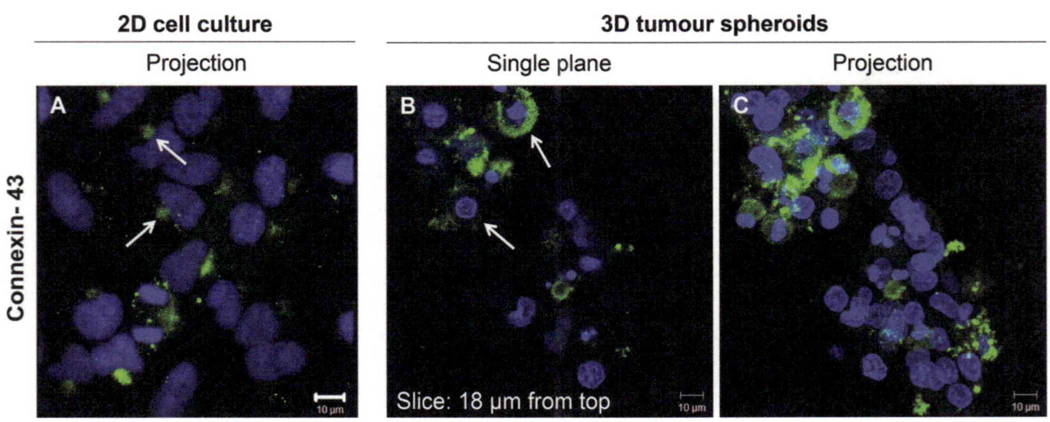

Figure 9. LSCM images of (A) 2D cell cultures and (B–C) 3D tumour spheroids stained with Alexa Fluor 488-conjugated Monoclonal Mouse Anti-Human Connexin-43 IgG (1:50 dilution) and Hoechst nuclear counterstain (1:400 dilution). 3D tumour spheroids were formed in ULA plates, without Happy Cell™ ASM. Arrows highlight (A) expression of connnexin-43 plaques in 2D cell cultures and (B) cortical expression of connexin-43 in 3D tumour spheroids. (C) Projection of Z-stack images were obtained with the projection function of Zeiss LSM5 software (46 sections, total height: 58.5 μm). (A–C) Scale bars: 10 μm (63× magnification).

4.2.1.2. *Advanced microscopy*

The protocol presented within this section contains a series of detailed steps that allow examining the morphology of cultured cells subsequent to 3D growth by advanced microscopy, such as, for example, scanning electron microscopy (SEM) or He-ion microscopy (HIM).

It is widely accepted that the preparation methods of biological specimens for SEM/HIM imaging can introduce artefacts. To avoid this and to preserve the cells' architecture, we recommend employing a drying method using hexamethyldisilazane (HMDS). HMDS represents a cost- and time-efficient alternative to critical point drying (CPD) in the preparation of cells for electron microscopy imaging [106, 107].

A. Dehydration of fixed 3D tumour spheroids:

- Following fixation, immerse samples in PBS to remove excess fixative for 10 min—all steps are carried out at room temperature.
- Gently aspirate the 3D tumour spheroids with a 1000 μL tip and transfer them on microscope glass slides previously marked with a liquid-repellent slide marker pen for staining procedures.

  Please note: the use of a 1000 μL tip is necessary to avoid perturbation of the original 3D spheroids architecture.
- Add an excess of 30% ethanol (EtOH) to samples, and leave them to equilibrate for 10 min.
- Discard excess liquid. Excess solutions should be discarded by gentle pipetting (avoid vacuum-pipetting) out the liquids, making sure that 3D tumour spheroids (visible by naked eye as small white spots after addition of EtOH) are not accidentally aspirated during this step.
- Repeat previous steps with 50, 70 and 90% EtOH, and leave to equilibrate for 10 min for each solution—discard solutions after each incubation.
- Add an excess of absolute EtOH and leave to equilibrate for 20 min—discard excess liquid. Repeat this step twice, discarding the solution after each incubation.

B. Chemical drying of dehydrated 3D tumour spheroids:

- Add an excess of 30% HDMS and leave to equilibrate for 5 min. Handle HDMS with care. HDMS is a flammable liquid and vapour. It is harmful if swallowed, inhaled or absorbed through skin. It causes severe irritation or burns to skin, eyes and respiratory tract. Avoid contact with eyes and skin. Avoid breathing vapours. Wear goggles, gloves and protective clothing. Use only with adequate ventilation. Wash hands thoroughly after handling. Avoid prolonged or repeated exposure.
- Discard excess solution.
- Add an excess of 60% HDMS and leave to equilibrate for 5 min.
- After discarding the solution, add an excess of pure HDMS and equilibrate for 10 min.
- Remove excess liquid and dry in a chemical fumehood overnight.

4.2.1.2.1. Experimental validation: imaging of 3D architecture, cell membrane texture and cell shape by HIM

He-ion microscopy (HIM) is an advanced imaging technique in which a focused beam of He^+ ions is directed onto the sample surface, which liberates secondary electrons that are collected forming detailed images of the sample surface topography [108]. In biomedical sciences, HIM offers various advantages over conventional SEM imaging, such as a high spatial resolution (with better material contrast and improved depth of focus) and the ability to image uncoated, non-conductive samples without the deposition of a metal (or other conductive) overcoat [109, 110], which can indeed reduce and/or completely mask cell surface details [111]. This opens up a whole new range of surface details in biological specimens that can be examined rapidly and with less risk of artefacts. HIM accuracy can be indeed exploited to image the shape, membrane texture, membranous projections and 3D architecture of 3D tumour spheroids with unsurpassed image quality and detail.

Our protocol for advanced microscopy was found valuable for capturing HIM images of 3D tumour spheroids formed without Happy Cell™ ASM in 96-well ULA plates (**Figure 10**).

4.2.2. Molecular analysis

Because of the central role of proteins in understanding cancer cells responses, it is often valuable to be able to extract, purify and quantify the expression of specific proteins. Thus, we developed an easy-to-use protocol outlining the steps necessary for the extraction of total protein from 3D tumour spheroids.

A. Collect 3D tumour spheroids:

- Aspirate growth media and 3D tumour spheroids formed with a 1000 μL pipette from each well and collect them in a 20 mL tube.

  Please note the use of a 1000 μL tip is necessary to avoid perturbation of the original 3D spheroids architecture

- Centrifuge at 5000 rpm at room temperature (25°C) for 20 min.

B. Total protein extraction:

- Aspirate the supernatant and re-suspend the cell-pellet with 1.5 mL of PBS.
- Centrifuge at 5000 rpm at room temperature (25°C) for 20 min.
- Aspirate the supernatant and re-suspend the cell-pellet in lysis buffer.
- After addition of the lysis buffer, mix energetically to favour disaggregation of 3D tumour spheroid-pellet.
- Proceed with standard immunoblotting.

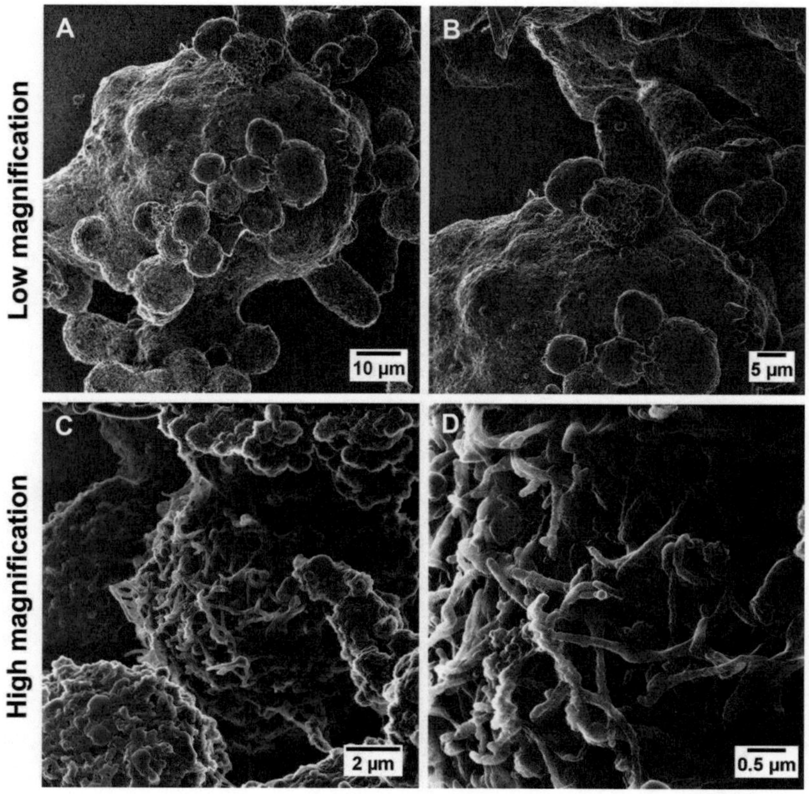

Figure 10. HIM images of A549 cells forming 3D tumour spheroids in ULA plates. The three-dimensional architecture of the spheroids can be appreciated at low magnification (A–B), while the presence of unique membrane ruffles and a multitude of filopodia and lamellipodia on the cells surface can be detected when imaging at high magnifications (C–D). Samples were imaged using an accelerating voltage of 30 kV. The working distance was 8 mm and a 10 µm aperture. The probe current was between 0.5 and 1.5 pA. Images were acquired by collecting the secondary electrons emitted by the interaction between the He-ion beam and the specimen with an Everhart-Thornley detector (part of the He-ion microscope system). The image signal was acquired in a 32- or 64-line integration to each contributing line of the image. Scale bars: (A) 10 µm, (B) 5 µm, (C) 2 µm and (D) 0.5 µm.

4.2.2.1. Experimental validation: Western Blot of cell junctions

Validation data were obtained using our protocol to extract protein from 3D tumour models cultured in Happy Cell™ ASM in Nunc® low-cell binding plates (24-well format), as confirmation of the molecular fingerprinting of the 3D tumour spheroids. Western blotting, a gold-standard technique in protein detection and quantification, was used for evaluating the expression in 3D tumour spheroids of (i) adherent junctions (namely, full-length protein E-cadherin and β-catenin) and (ii) TJs, such as occluding, zonula occludens-1 (ZO-1) and ZO-3. Cell lysis and Western blotting were performed following previously optimised in-house published protocols [112]. The results obtained (shown in **Figure 11**) demonstrated the expression of full-length E-cadherin and β-catenin. It should be noted that cells cultured in 3D spheroids showed nuclear translocation of cleaved E-cadherin (**Figure 7**) as well as

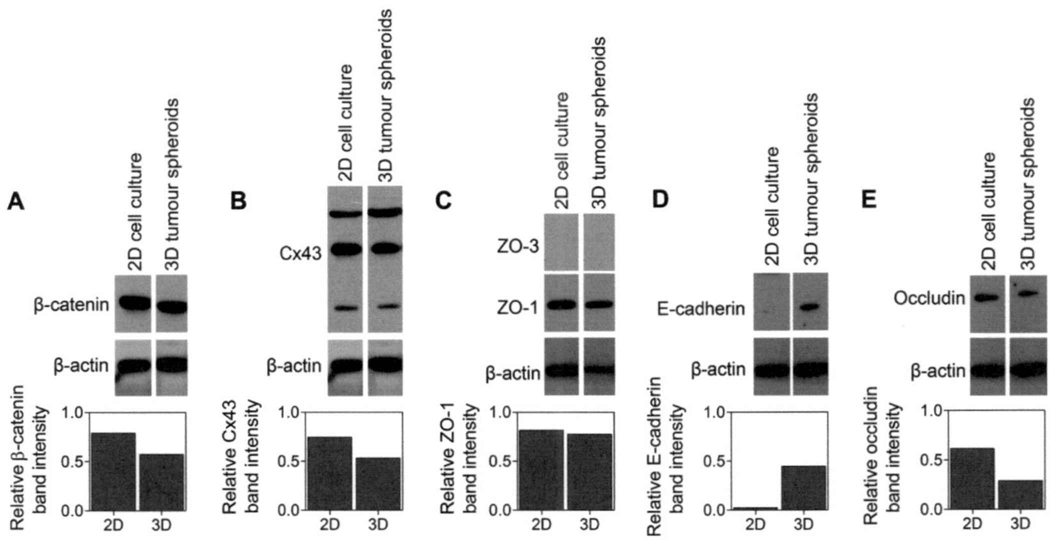

Figure 11. Western blot of (A) β-catenin (92 kDa), (B) Connexin-43 (19, 31 and 59 kDa), (C) ZO-3 (140 kDa) and ZO-1 (220 kDa), (D) full-length protein E-cadherin (135 kDa), (E) occludin (65 kDa), and relative bands intensities normalised on the β-actin bands intensity. 2D cultures and 3D tumour spheroids were lysed following the protocol described—the lysates were resolved by SDS-PAGE and after Western blotting were probed for the various proteins.

the expression of full-length E-cadherin (**Figure 11**): co-expression of these two forms of the E-cadherin protein is in fact possible during intermediate steps of the EMT process. In addition, Western blot showed that occluding and ZO-1 were expressed, whereas ZO-3 was not. For forming functional TJs, the transmembrane components of occludin need to interact with at least one ZO protein [113]. Our results seemed to suggest therefore that functional TJs might have been formed in 3D tumour spheroids.

4.3. Protocol for the evaluation of biological responses in 3D tumour spheroids

Exposure of 3D tumour spheroids formed with our protocol to drugs or nanomedicine candidates resulted very simple, as the spheroids could be exposed to the tested compounds/nanocarriers by their direct addition in the media, as conventionally used in 2D cell culturing protocols. In order to quantify the cellular responses to the drugs/nanocarriers tested, we developed protocols based on commercially available assays that are routinely applied to monitor intracellular activity in 2D monolayer cell cultures.

Various techniques are used to characterise the effect of anticancer agents on 3D tumour spheroids, as described in a recently published review [95]. In general, monitoring of spheroid integrity by phase contrast imaging is still the most popular technique to evaluate the cytotoxic effects of drugs. Spheroid growth delay, a classical analytical endpoint, is most frequently calculated as the difference between treated and untreated spheroids to reach a particular volume and has recently been proposed for standardised spheroid screening [114]. However, with respect to clinical relevance, loss of spheroid integrity may not result from total tumour cells destruction, as spheroid integrity does not necessarily reflect the presence of viable cells. Therefore, quantification of cell survival after treatment is desirable. Our protocols for conventional and high-throughput assays can be used for visualising and monitoring the viability of the cell population in the spheroids. Thus, they allow for the investigation of (i) the therapeutic efficacy of tested chemotherapeutic agents, (ii) the toxicity of nanocarriers and/or (iii) the efficiency in delivering anticancer treatments by means of nanocarriers.

4.3.1. Conventional assays

Standardised protocols, which do not require further optimization depending on the rate of cell growth or the cell culturing substrate used, are described in detail for the live/dead cytotoxicity and trypan blue exclusion assays in the following sections.

A. Live/dead cytotoxicity assay in 3D tumour spheroids:

- At the desired time point, harvest cells by gentle pipetting 100 μL/well media. This operation cannot be performed with vacuum pipettes, to avoid accidental aspiration of the 3D tumour spheroids formed.
- Add the pre-warmed staining solution (100 μL/well) to the wells. Staining solution is prepared as recommended in the supplier's protocol.
- Incubate for 45 min at room temperature in the dark.
- Gently aspirate the 3D tumour spheroids with a 1000 μL tip and proceed to analysis. Please note: the use of a 1000 μL tip is necessary to avoid any potential perturbation in the spatial localization of live/dead cells.
- Please note that it is not possible to carry out analysis on stored specimens.

B. Trypan blue exclusion assay in 3D tumour spheroids:

- Aspirate growth media and 3D tumour spheroids from each well with a 1000 μL pipette and collect them in a 20 mL tube. Please note the use of a 1000 μL tip is necessary to avoid perturbation of the original 3D spheroids architecture, triggering any potential cell death.
- Centrifuge at 3000 rpm at room temperature (25°C) for 5 min.
- Aspirate the supernatant and re-suspend the cells with 1 mL of PBS.
- Add 100 μL of trypan blue solution (0.4%) to 1 mL of cells.
- After addition of the trypan blue solution, mix energetically to favour disaggregation of 3D tumour spheroids.
- Load a haemocytometer or a cell counter slide with the stained cell suspension.
- Immediately count the number of blue stained cells and the number of unstained cells under a low magnification microscope or with an automatic cell counter.
- To calculate the percentage (%) of live cells use Eq. (1):

$$\% \text{ live cells} = \left[1 - \frac{\text{number of stained cells}}{\text{number of total cells}}\right] \times 100 \quad (1)$$

4.3.1.1. Experimental validation: therapeutic efficacy of Paclitaxel in 3D tumour spheroids

In the validation of our protocol, the live/dead cytotoxicity assay was integrated with LSCM. 3D tumour spheroids were formed without Happy Cell™ ASM in 96-well ULA plates. 3D tumour spheroids were then exposed to Paclitaxel (0.03 mM) for 24 and 72 h, and the percentage of live cells was compared to the paired negative (untreated cells) and positive controls (cells treated with 70% MeOH for 30 min at 37°C). Representative LSCM images of 3D tumour spheroids stained with the devised protocol for live/dead cytotoxicity assay are shown in **Figure 12**. Analysis of LSCM images demonstrates a small (but not significant) increase in the proportion of dead cells when 3D tumour spheroids were exposed to Paclitaxel for 24 h, while the size of the spheroids appeared reduced following exposure at both time points. The treatment with 70% MeOH provided the expected positive control outcome.

Further to this, the viability of 3D tumour spheroids exposed to Paclitaxel (0.03 mM) for 24 and 72 h was evaluated by means of the trypan blue exclusion assay protocol. Quantitative data are reported in **Figure 13** as the average ± standard deviation ($n_{replicates} \geq 2$). Exposure to Paclitaxel for 24 h resulted in a significant reduction in cell viability in 3D tumour spheroids, although cell viability increased after 72 h treatment. This result might be related to the well-known ability of 3D tumour spheroids to mimic drug resistance phenomena [115, 116], which has been observed *in vivo*. As expected, treatment with 70% MeOH resulted in a significant reduction of live cells as positive control sample.

4.3.1.2. Experimental validation: cytotoxicity of nanocarriers in 3D tumour spheroids and therapeutic efficacy of anticancer agents delivered by nanocarriers

3D tumour spheroids were formed without Happy Cell™ ASM in 96-well ULA plates and exposed to gold nanoboxes (Au-1) and its nanomedicine form consisting of a multi-layered nanomaterial made of the same gold core (Au-1) coated with Paclitaxel-encapsulated gelatine (Au-2). Full characterisation of these nanomaterials is provided in a previous publication of the authors [117]. 3D tumour spheroids were exposed to Au-1 and Au-2 for 24 or 72 h. The experimental design included a negative control (untreated cells) and a positive control (cells treated with 70% MeOH for 30 min at 37°C). Representative LSCM images of 3D tumour spheroids stained with the live/dead cytotoxicity assay protocol are shown in **Figure 14**. Analysis of such images allowed concluding that Au-1 did not produce any significant variation in the cell viability following 24 and 72 h exposure, whereas 72 h exposure to Au-2 resulted in a reduction of the total amount of live cells in the 3D tumour spheroids.

Trypan blue exclusion assay was carried out in parallel experiments, where 3D tumour spheroids formed in 96-well ULA plates were exposed to Au-1 for 24 and 72 h. The quantitative results are reported in **Figure 15** as average ± standard deviation ($n_{replicates} \geq 2$). The assay confirmed that such nanocarriers did not cause any cytotoxicity when incubated with 3D tumour spheroids for 72 h.

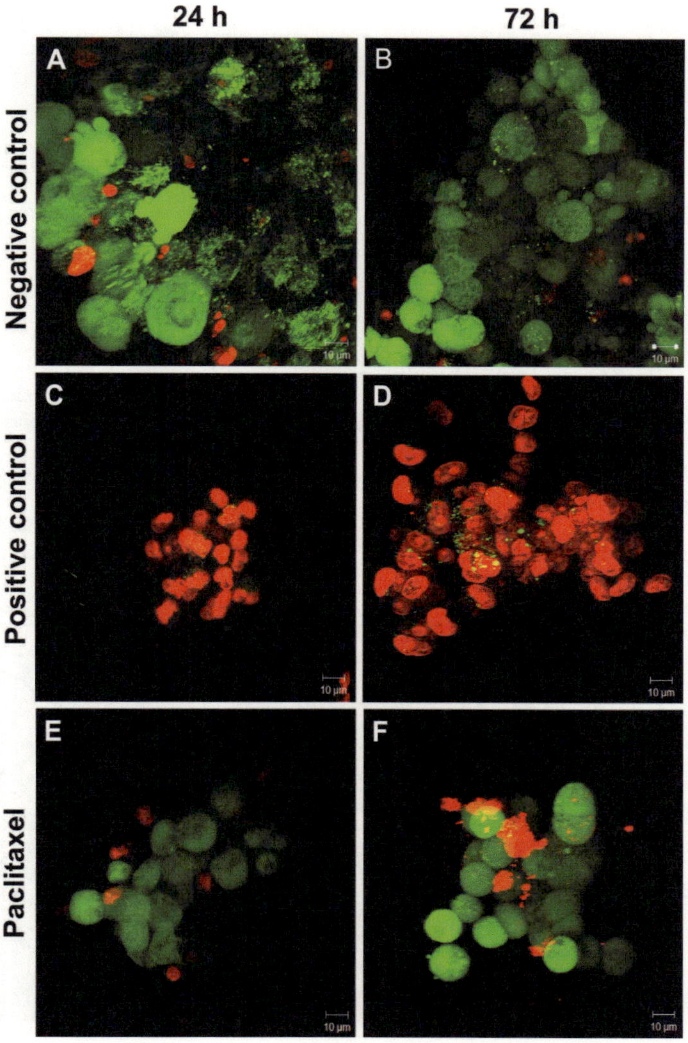

Figure 12. LSCM images of 3D tumour spheroids stained for live (calcein AM; 1:500 dilution; in green) and dead (ethidium homodimer-1; 1:300 dilution; in red) cells after 24 and 72 h exposure to Paclitaxel (0.03 mM). Negative (untreated 3D tumour spheroids) and positive (3D tumour spheroids treated with 70% MeOH for 30 min) controls are also reported. Scale bars: 10 μm (63× objective lens).

4.3.2. High-throughput assays

The setup based on high-throughput flow cytometry is described below. This setup could be easily integrated into a standard large-scale drug-testing routine. The setup requires a small number of spheroids and a limited amount of drug/nanomedicine candidate, and it can integrate many analytical endpoints. Since spheroid dissociation is carried out during the sample preparation steps, this assay is suitable for all spheroid size ranges that may be examined when comparing treated and untreated spheroids.

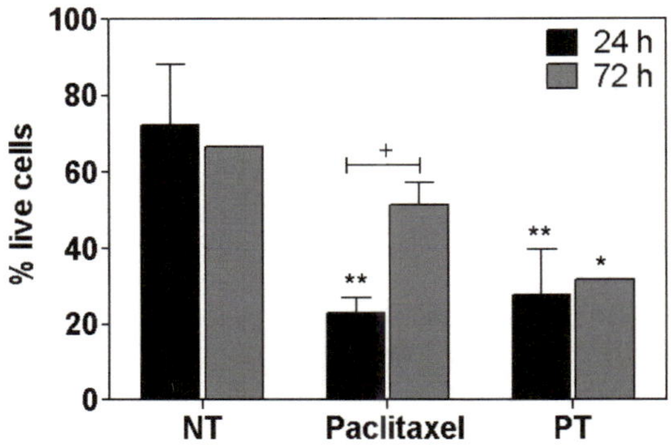

Figure 13. Percentage (%) of live cells in 3D tumour spheroids grown in ULA plates without Happy Cell™ ASM and exposed to Paclitaxel (0.03 mM) for 24 and 72 h. Untreated (negative control or NT) and 70% MeOH-treated (positive control or PT) 3D tumour spheroids were also included in the experimental design. Stained and unstained cells were counted by a Countess™ cell counter. For the statistical analysis, a two-way analysis of variance (2-way ANOVA) followed by Bonferroni post-test analysis was carried out (GraphPad Prism 5 Software Inc., USA). $p < 0.05$ was considered statistically significant. The symbols (*) and (**) indicate $p < 0.05$ and $p < 0.01$, as compared to NT. The symbol (+) indicates $p < 0.05$ as compared to 24 h.

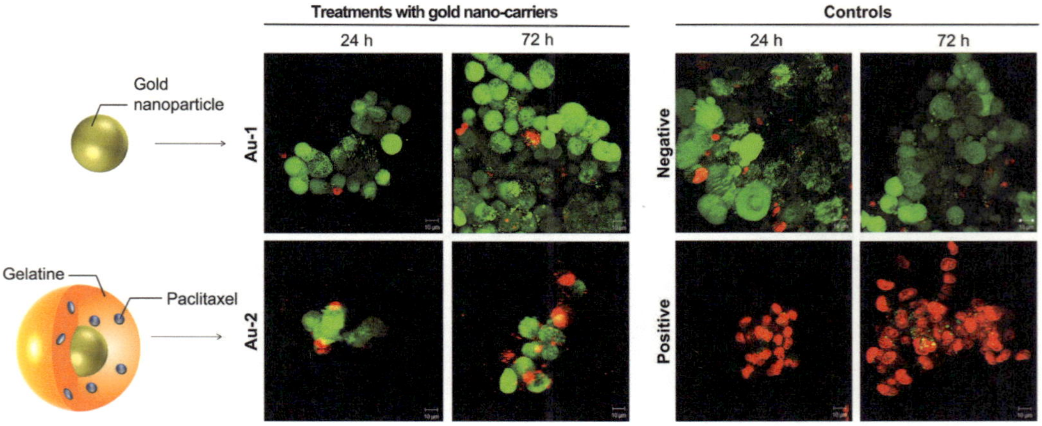

Figure 14. LSCM images of 3D tumour spheroids stained for live (calcein AM; 1:500 dilution; in green) and dead (ethidium homodimer-1; 1:300 dilution; in red) cells after 24 and 72 h exposure to a gold nanocarrier (Au-1) or Paclitaxel-loaded, gelatine-coated gold nanocarriers (Au-2). Negative (untreated 3D tumour spheroids) and positive (3D tumour spheroids treated with 70% MeOH for 30 min) controls are also reported. Scale bars: 10 μm (63× objective lens).

A. Sample preparation:

- At the desired time point, aspirate growth media and 3D tumour spheroids from each well with a 1000 μL pipette and collect them in a 20 mL tube.
- Centrifuge at 1650 rpm at room temperature (25°C) for 5 min.

B. Assay:

- Perform the cell staining with the kit chosen according to the manufacturer's protocol, and run the sample in the flow cytometer.

- Special considerations: please note that the assay kit used in this study does not require a step for disaggregating the 3D tumour spheroids in a single cell suspension, as lysis of cells is performed during the assay. An additional step might be required if a different kit is used.

4.3.2.1. Experimental validation: changes in cell cycle following exposure to Paclitaxel

Example data were obtained using the protocol described above to quantify the cell cycle changes in 3D tumour spheroids exposed to Paclitaxel (0.03 mM) for 24 h. Stained nuclei were visualised using the SSC-H vs FSC-H scatter plot and a gate (P1) was applied to exclude debris at lower scatter intensities. Aggregate exclusion gating (P2 in P1) via doublet discrimination was performed on the P1 population using the FL2-H vs FL2-A scatter. Finally, analysis of the cell cycle stage for G0/G1, S and G2/M phase was carried out by manual gating on the FL2-H histogram. A minimum of 10,000 events was collected in the (P2 in P1) gate and visualised on the FL2-H histogram. Data are presented in **Figure 16** as percentage (%) cell population in (P2 in P1) and expressed as average ± standard deviation ($n_{test} = 2$).

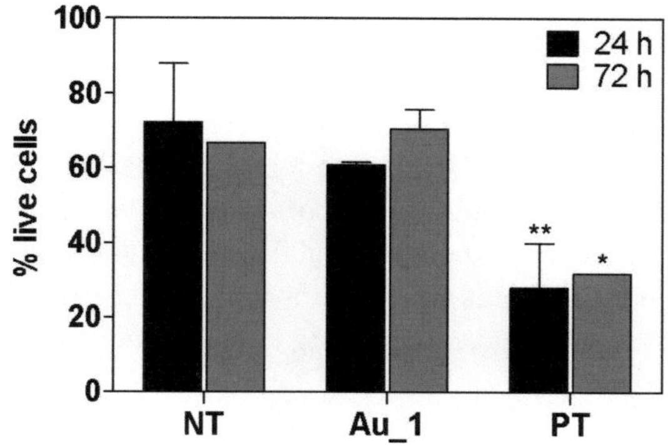

Figure 15. Percentage (%) of live cells in 3D tumour spheroids grown in ULA plates after exposure to gold nanoparticles for 24 and 72 h. Untreated (negative control or NT) and 70% MeOH-treated (positive control or PT) 3D tumour spheroids were also included in the experimental design. Stained and unstained cells were counted by a Countess™ cell counter. For the statistical analysis, a two-way ANOVA followed by Bonferroni post-test analysis was carried out (GraphPad Prism 5 Software Inc., USA). $p < 0.05$ was considered statistically significant. The symbols (*) and (**) indicate $p < 0.05$ and $p < 0.01$ as compared to NT.

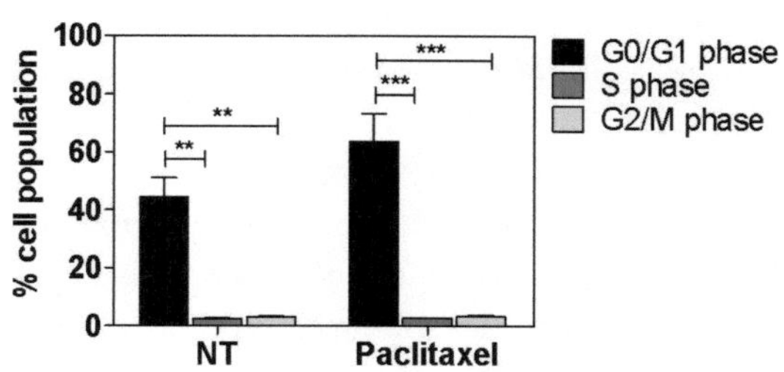

Figure 16. Percentage (%) of cell population in the G0/G1, S and G2/M phases after exposure to Paclitaxel (0.03 mM) for 24 h. cells were stained with the BD Cycletest™ Plus DNA Reagent Kit and analysed using BD Accuri™ C6 flow cytometer. Data are shown as average ± standard deviation ($n_{replicates} = 2$). For the statistical analysis, a two-way ANOVA followed by Bonferroni post-test analysis was carried out (GraphPad Prism 5 Software Inc., USA). $p < 0.05$ was considered statistically significant. The symbols (**) and (***) indicate significant changes ($p < 0.01$ and $p < 0.001$, respectively) compared to the G0/G1 phase.

Our data showed an increase in the percentage cell population in G0/G1 phase after 24 h treatment with Paclitaxel. This anticancer drug is known to inhibit the mitotic spindle formation, thus blocking the progression of mitosis and triggering apoptosis or reversion to the G-phase of the cell cycle without cell division. The results presented in **Figure 16** reflect the mechanism of action of the drug under investigation, and validate the trypan blue exclusion assay results, showing an increased cell death following 24 h exposure to Paclitaxel (**Figure 13**).

5. Authors' perspective

Since 2006, when a simple method to generate non-adherent 3D tumour spheroids for potential high-throughput toxicity analysis of drug compounds was firstly reported [118], several methods have been developed to generate scaffold-free 3D tumour spheroids for drug discovery. These methods share the common feature of promoting cell-cell coupling by resisting cell-substrate interactions. They can be grouped into four main methodological categories: (i) hanging drop [85, 119, 120]; (ii) cell culture on non-adherent surfaces that effectively inhibits cellular attachment, such as (poly-HEMA)-coated plates [121, 122], low-binding plates [123, 124], ultra-low attachment (ULA) plates [125] or micro-patterned plates [126, 127]; (iii) microcarrier systems [128] and (iv) rotation-based culturing techniques, such as spinner flasks [129] and rotary cell culture systems [130, 131]. Formation of scaffold-free heterotypic multicellular 3D spheroids has likewise been extensively reported [132]. Nevertheless, when transferring drug testing to the third dimension, the limited ability to image, process and automate assays performance in 3D tumour spheroids still remains the main barrier for the adoption of these 3D culture techniques for routine preclinical drug development studies [133].

To the best of our knowledge, the main commercial approaches to the automation of scaffold-free 3D cell culturing techniques are InSphero GravityPLUS™ Hanging Drop system and GravityTRAP™ ULA plates, 3D Biomatrix's Perfecta3D™ hanging drop multi-well plates, Global Cell Solutions magnetic microcarrier-based GEM™ and PolyGEM™ systems, and Happy Cell™ ASM. Although each method has certain advantages, the challenges that these approaches still pose for automation are well-documented [24]. In detail, limitations are posed to imaging due to Z-axis resolution, image depth and light scattering [134], although advances have been recently made in this field [135] by, for example, introducing additional post-processing treatments of 3D spheroids with Scale reagent [136]. The main consequence is that most studies on 3D tumour spheroids are carried out at low resolution by light microscopy and immunohistochemistry, or at the single-cell level by multiphoton microscopy or spectroscopic imaging/mapping techniques [137]. These techniques however are not compatible with high-throughput screening. In addition to these technical aspects, most of the commercially available products for scaffold-free 3D cell culture are not accompanied by technical documentation or peer-reviewed scientific publications describing the specific protocols that need to be used for drug discovery applications. For many of the products mentioned above, users and operators are left with the burden of re-optimising the working protocols to their specific needs. For instance, it is known that cells in Happy Cell™ ASM, a product used in this study, form discrete populations within the wells of 96-well low-cell binding plates provided by the same supplier and, during continued culture, they assemble into spheroids. However, no information is provided on: (i) the size of the 3D tumour spheroids formed in the multi-well plates supplied by Biocroi Ltd itself; (ii) the size of the 3D tumour spheroids obtained if using multi-well plates purchased from different commercial sources and compatible with high-throughput screening assays and (iii) how to standardise and homogenise the size of 3D tumour spheroids. Similarly, according to the manufacturer, protocols for the formation, maintenance and harvesting of 3D tumour spheroids formed in 3D Biomatrix's Perfecta3D™ hanging drop multi-well plates should be optimised by each user[138–140]. Such uncertainties are indeed strongly hindering the integration of 3D culturing technologies into the pharmaceutical industry practice.

The aim of this study was to develop practical and technical solutions filling such gaps of standardisation, focusing on screening applications that are achievable without expensive robotic equipment and that could be employed in many research labs. Happy Cell™ ASM was selected as the least characterised product, whereas low-binding/ULA multi-well plates were chosen as they are the most common substrates currently used for forming 3D tumour spheroids. Human adenocarcinoma cells (A549 cell line) were selected as a relevant model for developing technical protocols with translational potential in the drug development pipeline: all results presented here are based on such cell line. The A549 cell line is considered one of the closest cell line mimicking the human epithelial alveolar model [141], and a physiologically relevant *in vitro* model of NSCLC [142], which is the most prevalent form of human lung cancer originating from epithelial cells. A549 cells are also an established cell line for anticancer drugs discovery and screening [143] and it has proven to be a robust cell line/alveolar model for several previous nanomaterial-based studies [112, 117, 144–148]. Unlike other studies on A549 cells [118, 149–151], the formation of our 3D tumour spheroids did not require architecturally complex scaffolds or reconstitute basement membrane that could circumvent some applications, such as drug delivery and assays performance.

The protocols developed can be grouped in three categories (in dark blue in **Figure 1**). For each of the identified categories, experimental validation was carried out and is reported in the form in which they have been used in our lab, namely forming 3D tumour spheroids and incorporating them into a 3D co-culture model, characterising the markers expressed by 3D tumour spheroids, and testing an anticancer drug or gold nanocarriers.

Our protocol allowed cell aggregation and spheroids formation in standard commercially available multi-well plates, thus meeting the pharmaceutical industry requirements [133]. Formation of 3D tumour spheroids was easily controlled by tuning specific parameters (i.e. the initial cell aggregation and cells adhesion to the substrate). 3D structure and mechanical robustness are mandatory requirements to allow the translation and implementation of 3D tumour spheroids into a primary drug-testing routine. Without these features, in fact, the pathophysiological gradients developed in genuine tumour spheroids cannot be mimicked [94]. Unfortunately, some so-called spheroids in the literature are no more than loose aggregates that easily detach; they cannot be manipulated or transferred and they show lack cell-cell interactions [23]. Our experimental data demonstrated that only 3D tumour spheroids formed in 24-well low-cell binding plates and ULA plates were mechanically robust (**Figure 3**). 3D tumour spheroids formed in these cell culture plates could be easily transferred onto different substrates for further specific analysis (e.g. immunofluorescence staining and imaging), showing a beneficial advantage over other multi-well cell plates.

As heterotypic cell interactions are key for the function of certain tissues, co-cultures including multiple cell types are another means of increasing relevance to *in vivo* scenario [152]. A protocol for forming 3D co-culture tumour models was therefore developed. Various examples of spheroid co-culture approaches are reported in the literature [153] including, but not limited to: (i) mixed spheroids, mimicking, for example, lung cancer [154]; (ii) tumour spheroids cultured on fibroblast monolayers as model of colon carcinoma [155, 156] or breast cancer [157] and (iii) tumour spheroids co-cultured with pre-established fibroblast spheroids for creating *in vitro* models of breast cancer [158–160]. Mixed spheroids are the most widely used approach. In this approach, however, the histo-morphology and cellular distribution cannot be accurately controlled by the end-user, resulting in not-homogenous (and not replicable) models, even if identical cell numbers and harmonised culture protocols are used. These models have therefore limited application as drug screening models in the pharmaceutical industry. Thus, in our study a co-culture approach in which tumour spheroids were plated onto a fibroblast monolayer was preferred. Fibroblasts were selected as these cells represent a major portion of the tumour stroma in carcinomas, and they can promote cancer progression and invasion [161, 162]. Confocal images of the 3D co-culture models formed in this study demonstrated the successful attachment of 3D spheroids of A549 cells to the fibroblast monolayer (**Figure 4**), suggesting interaction among the different cell types.

Since the presence or extent of certain tumour characteristics, such as those associated to EMT, are likely to affect drug response, it is essential to extensively characterise the spheroid model used during drug screening. The protocols described in our study allowed a simple and economically feasible evaluation of cell markers expression (**Figures 5–10**) by means of conventional and advanced microscopy approaches. Spheroids characteristics were also assessed by Western blotting (**Figure 11**). Although not all the main features of solid cancers (such as the influence of acellular stroma and immune cells) were modelled through our formation protocols, our results demonstrated that 3D tumour spheroids did reflect many important properties of solid tumours, including the expression of specific cancer markers.

Finally, keeping in mind that one of the preferred embodiments of the protocols described herein is their application in research labs for efficacy screening of new chemotherapeutic agents and oncological nanomedicine products, cell features/responses were assessed by conventional and high-throughput assays (**Figures 12–16**). Various studies provide evidence that standard cell proliferation assays (e.g. AlamarBlue assay [163]) are not suitable for quantifying cytotoxic/cytostatic responses in 3D tumour spheroids, as they do not show a linear correlation with cell densities [164], unless properly adapted. Similarly, nanomaterials are known to interfere with many assays (e.g. MTT) [3, 165]. Our protocols based on conventional viability assays satisfied criteria, such as ease-of-use and reliability, without losing the main advantages of validated commercially available kits. A protocol for high-throughput flow cytometry was also developed, resolving time and cost issues [166]. The implementation of flow cytometry can be regarded as advantageous since this technique is widely used in clinical laboratories and can include multiple molecular read-outs and endpoints.

http://dx.doi.org/10.5772/intechopen.69447

6. Conclusions

3D tumour spheroids can allow the fast development of new effective anticancer nanomedicines without dissatisfying safety, economical and ethical issues raised by societal, healthcare and pharmaceutical industry stakeholders. However, this can be achieved only by developing clear, simple and reproducible protocols for the formation and use of 3D tumour spheroids in a way that is compatible with standard analysis techniques. The protocols described within this work offer a set of tools addressing many of the problems that need to be overcome in order to translate 3D tumour spheroids into preclinical models. The obtained data validate the developed protocols to bear a novel platform technology solving the existing bottlenecks in the arena of anticancer drug and nanomedicine products development, presented here in the form of gold nanocarriers.

Acknowledgements

This work was partially supported by the CRANN Institute (CRANN Pathfinder to DM), the EU FP7 NAMDIATREAM project (NMP-2009-LARGE-3-246479) and the Irish Research Council (IRC) Government of Ireland Postdoctoral Fellowship to DM.

The authors would like to thank Biocroi Ltd for the materials supplied free-of-charge, and Kieran Crosbie Staunton for Western blotting.

Conflict of interest

The authors declare that they have no conflict of interest.

Author details

Dania Movia[1,2] and Adriele Prina-Mello[1,2,3,4]*

*Address all correspondence to: prinamea@tcd.ie

1 Laboratory for Biological Characterization of Advanced Materials (LBCAM), School of Medicine, Trinity Translational Medicine Institute (TTMI), Trinity College, Dublin, Ireland

2 Department of Clinical Medicine, School of Medicine, Trinity Translational Medicine Institute (TTMI), Trinity College, Dublin, Ireland

3 CRANN Institute, University of Dublin, Trinity College, Dublin, Ireland

4 AMBER Centre, University of Dublin, Trinity College, Dublin, Ireland

References

[1] Shi J, et al. Cancer nanomedicine: Progress, challenges and opportunities. Nature Reviews Cancer. 2017;**17**(1):20-37. DOI: 10.1038/nrc.2016.108

[2] Moffat JG, Rudolph J, Bailey D. Phenotypic screening in cancer drug discovery—Past, present and future. Nature Reviews Drug Discovery. 2014;**13**(8):588-602. DOI: 10.1038/nrd4366

[3] Bregoli L, et al. Nanomedicine applied to translational oncology: A future perspective on cancer treatment. Nanomedicine. 2016;**12**(1):81-103. DOI: 10.1016/j.nano.2015.08.006

[4] Wicki A, et al. Nanomedicine in cancer therapy: Challenges, opportunities, and clinical applications. Journal of Controlled Release. 2015;**200**:138-157. DOI: 10.1016/j.jconrel.2014.12.030

[5] Hartung T. Look back in anger—What clinical studies tell us about preclinical work. ALTEX. 2013;**30**(3):275-291

[6] Hakanson M, Cukierman E, Charnley M. Miniaturized pre-clinical cancer models as research and diagnostic tools. Advanced Drug Delivery Reviews.. 2014;**69-70**:52-66. DOI: 10.1016/j.addr.2013.11.010

[7] Kola I, Landis J. Can the pharmaceutical industry reduce attrition rates? Nature Reviews Drug Discovery. 2004;**3**(8):711-715. DOI: 10.1038/nrd1470

[8] Arrowsmith J, Miller P. Trial watch: Phase II and phase III attrition rates 2011-2012. Nature Reviews Drug Discovery. 2013;**12**(8):569. DOI: 10.1038/nrd4090

[9] Hay M, et al. Clinical development success rates for investigational drugs. Nature Biotechnology. 2014;**32**(1):40-51. DOI: 10.1038/nbt.2786

[10] Arrowsmith J. Trial watch: Phase II failures: 2008-2010. Nature Reviews Drug Discovery. 2011;**10**(5):328-329. DOI: 10.1038/nrd3439

[11] Venditto VJ, Szoka FC Jr. Cancer nanomedicines: So many papers and so few drugs! Advanced Drug Delivery Reviews. 2013;**65**(1):80-88. DOI: 10.1016/j.addr.2012.09.038

[12] Kamb A. What's wrong with our cancer models? Nature Reviews Drug Discovery. 2005;**4**(2):161-165. DOI: 10.1038/nrd1635

[13] Unger C, et al. Modeling human carcinomas: Physiologically relevant 3D models to improve anti-cancer drug development. Advanced Drug Delivery Reviews. 2014;**79-80**:50-67. DOI: 10.1016/j.addr.2014.10.015

[14] Cook D, et al. Lessons learned from the fate of AstraZeneca's drug pipeline: A five-dimensional framework. Nature Reviews Drug Discovery. 2014;**13**(6):419-431. DOI: 10.1038/nrd4309

[15] Pound P, et al. Where is the evidence that animal research benefits humans? BMJ. 2004;**328**(7438):514-517. DOI: 10.1136/bmj.328.7438.514

[16] Hartung T. Food for thought… on animal tests. ALTEX. 2008;**25**(1):3-16

[17] Mak IW, Evaniew N, and Ghert M. Lost in translation: Animal models and clinical trials in cancer treatment. American Journal of Translational Research. 2014;**6**(2):114-118

[18] Olson H, et al. Concordance of the toxicity of pharmaceuticals in humans and in animals. Regulatory Toxicology and Pharmacology. 2000;**32**(1):56-67. DOI: 10.1006/rtph.2000.1399

[19] van der Worp HB, et al. Can animal models of disease reliably inform human studies? PLoS Medicine. 2010;**7**(3):e1000245. DOI: 10.1371/journal.pmed.1000245

[20] Hackam DG. Translating animal research into clinical benefit. BMJ. 2007;**334**(7586):163-164. DOI: 10.1136/bmj.39104.362951.80

[21] Perel P, et al. Comparison of treatment effects between animal experiments and clinical trials: Systematic review. BMJ. 2007;**334**(7586):197. DOI: 10.1136/bmj.39048.407928.BE

[22] Gillet JP, Varma S, Gottesman MM. The clinical relevance of cancer cell lines. Journal of the National Cancer Institute. 2013;**105**(7):452-458. DOI: 10.1093/jnci/djt007

[23] Weiswald LB, Bellet D, Dangles-Marie V. Spherical cancer models in tumor biology. Neoplasia. 2015;**17**(1):1-15. DOI: 10.1016/j.neo.2014.12.004

[24] Thoma CR, et al. 3D cell culture systems modeling tumor growth determinants in cancer target discovery. Advanced Drug Delivery Reviews. 2014;**69-70**:29-41. DOI: 10.1016/j.addr.2014.03.001

[25] Yamada KM, Cukierman E. Modeling tissue morphogenesis and cancer in 3D. Cell. 2007;**130**(4):601-610. DOI: 10.1016/j.cell.2007.08.006

[26] Nath S, Devi GR. Three-dimensional culture systems in cancer research: Focus on tumor spheroid model. Pharmacology and Therapeutics. 2016;**163**:94-108. DOI: 10.1016/j.pharmthera.2016.03.013

[27] Hickman JA, et al. Three-dimensional models of cancer for pharmacology and cancer cell biology: Capturing tumor complexity in vitro/ex vivo. Biotechnology Journal. 2014;**9**(9):1115-1128. DOI: 10.1002/biot.201300492

[28] Imamura Y, et al. Comparison of 2D- and 3D-culture models as drug-testing platforms in breast cancer. Oncology Reports. 2015;**33**(4):1837-1843. DOI: 10.3892/or.2015.3767

[29] Fitzgerald KA, et al. Life in 3D is never flat: 3D models to optimise drug delivery. Journal of Controlled Release. 2015;**215**:39-54. DOI: 10.1016/j.jconrel.2015.07.020

[30] Tay CY, et al. Reality check for nanomaterial-mediated therapy with 3D biomimetic culture systems. Advanced Functional Materials. 2016;**26**(23):4046-4065. DOI: 10.1002/adfm.201600476

[31] Wenzel C, et al. 3D high-content screening for the identification of compounds that target cells in dormant tumor spheroid regions. Experimental Cell Research. 2014;**323**(1):131-143. DOI: 10.1016/j.yexcr.2014.01.017

[32] Howes AL, et al. The phosphatidylinositol 3-kinase inhibitor, PX-866, is a potent inhibitor of cancer cell motility and growth in three-dimensional cultures. Molecular Cancer Therapeutics. 2007;**6**(9):2505-2514. DOI: 10.1158/1535-7163.MCT-06-0698

[33] Roberts DL, et al. Contribution of HIF-1 and drug penetrance to oxaliplatin resistance in hypoxic colorectal cancer cells. British Journal of Cancer. 2009;**101**(8):1290-1297. DOI: 10.1038/sj.bjc.6605311

[34] Dufau I, et al. Multicellular tumor spheroid model to evaluate spatio-temporal dynamics effect of chemotherapeutics: Application to the gemcitabine/CHK1 inhibitor combination in pancreatic cancer. BMC Cancer. 2012;**12**:15. DOI: 10.1186/1471-2407-12-15

[35] Nederman T. Effects of vinblastine and 5-fluorouracil on human glioma and thyroid cancer cell monolayers and spheroids. Cancer Research. 1984;**44**(1):254-258

[36] Carlsson J, Nederman T. A method to measure the radio and chemosensitivity of human spheroids. Advances in Experimental Medicine and Biology. 1983;**159**:399-417

[37] Acker H. The use of human tumor cells grown in multicellular spheroid culture for designing and improving therapeutic strategies. Journal of Theoretical Medicine. 1998;**1**:193-207

[38] Kunz-Schughart LA, Kreutz M, Knuechel R. Multicellular spheroids: A three-dimensional in vitro culture system to study tumour biology. International Journal of Experimental Pathology. 1998;**79**(1):1-23

[39] Kobayashi H, et al. Acquired multicellular-mediated resistance to alkylating agents in cancer. Proceedings of the National Academy of Sciences of the United States of America. 1993;**90**(8):3294-3298

[40] Graham CH, et al. Rapid acquisition of multicellular drug resistance after a single exposure of mammary tumor cells to antitumor alkylating agents. Journal of the National Cancer Institute. 1994;**86**(13):975-982

[41] Kerbel RS, et al. Induction and reversal of cell adhesion-dependent multicellular drug resistance in solid breast tumors. Human Cell. 1996;**9**(4):257-264

[42] Desoize B, Jardillier J. Multicellular resistance: A paradigm for clinical resistance? Critical Reviews in Oncology/Hematology. 2000;**36**(2-3):193-207

[43] Chatzinikolaidou M. Cell spheroids: the new frontiers in in vitro models for cancer drug validation. Drug Discovery Today. 2016;**21**(9):1553-1560. DOI: 10.1016/j.drudis.2016.06.024

[44] Rodriguez-Enriquez S, et al. Energy metabolism transition in multi-cellular human tumor spheroids. Journal of Cellular Physiology. 2008;**216**(1):189-197. DOI: 10.1002/jcp.21392

[45] Fallica B, Makin G, Zaman MH. Bioengineering approaches to study multidrug resistance in tumor cells. Integrative Biology. 2011;**3**(5):529-539. DOI: 10.1039/c0ib00142b

[46] Aguirre-Ghiso JA. Models, mechanisms and clinical evidence for cancer dormancy. Nature Reviews Cancer. 2007;**7**(11):834-846. DOI: 10.1038/nrc2256

[47] Ruppender NS, et al. Dormancy in solid tumors: Implications for prostate cancer. Cancer and Metastasis Reviews. 2013;**32**(3-4):501-509. DOI: 10.1007/s10555-013-9422-z

[48] Kyle AH, Baker JH, Minchinton AI. Targeting quiescent tumor cells via oxygen and IGF-I supplementation. Cancer Research. 2012;**72**(3):801-809. DOI: 10.1158/0008-5472.CAN-11-3059

[49] Ma HL, et al. Multicellular tumor spheroids as an in vivo-like tumor model for three-dimensional imaging of chemotherapeutic and nano material cellular penetration. Molecular Imaging. 2012;**11**(6):487-498

[50] Amann A, et al. Development of an innovative 3D cell culture system to study tumour-stroma interactions in non-small cell lung cancer cells. PLoS One. 2014;**9**(3):e92511. DOI: 10.1371/journal.pone.0092511

[51] Dolznig H, et al. Organotypic spheroid cultures to study tumor-stroma interaction during cancer development. Drug Discovery Today: Disease Models. 2011;**8**(2-3):113-119. DOI: 10.1016/j.ddmod.2011.06.003

[52] Griffin JL, Shockcor JP. Metabolic profiles of cancer cells. Nature Reviews Cancer. 2004;**4**(7):551-561. DOI: 10.1038/nrc1390

[53] Landowski TH, et al. Cell adhesion-mediated drug resistance (CAM-DR) is associated with activation of NF-kappa B (RelB/p50) in myeloma cells. Oncogene. 2003;**22**(16):2417-2421. DOI: 10.1038/sj.onc.1206315

[54] Serebriiskii I, et al. Fibroblast-derived 3D matrix differentially regulates the growth and drug-responsiveness of human cancer cells. Matrix Biology. 2008;**27**(6):573-585. DOI: 10.1016/j.matbio.2008.02.008

[55] Shield K, et al. Multicellular spheroids in ovarian cancer metastases: Biology and pathology. Gynecologic Oncology. 2009;**113**(1):143-148. DOI: 10.1016/j.ygyno.2008.11.032

[56] Eke I, Cordes N. Radiobiology goes 3D: How ECM and cell morphology impact on cell survival after irradiation. Radiotherapy and Oncology. 2011;**99**(3):271-278. DOI: 10.1016/j.radonc.2011.06.007

[57] Dalton WS. The tumor microenvironment as a determinant of drug response and resistance. Drug Resistance Updates. 1999;**2**(5):285-288. DOI: 10.1054/drup.1999.0097

[58] Leong DT, Ng KW. Probing the relevance of 3D cancer models in nanomedicine research. Advanced Drug Delivery Reviews.. 2014;**79-80**:95-106. DOI: 10.1016/j.addr.2014.06.007

[59] El-Dakdouki MH, Pure E, Huang X. Development of drug loaded nanoparticles for tumor targeting. Part 2: Enhancement of tumor penetration through receptor mediated transcytosis in 3D tumor models. Nanoscale. 2013;**5**(9):3904-3911. DOI: 10.1039/c3nr90022c

[60] Hu Q, et al. F3 peptide-functionalized PEG-PLA nanoparticles co-administrated with tLyp-1 peptide for anti-glioma drug delivery. Biomaterials. 2013;**34**(4):1135-1145. DOI: 10.1016/j.biomaterials.2012.10.048

[61] Xin H, et al. Anti-glioblastoma efficacy and safety of paclitaxel-loading Angiopep-conjugated dual targeting PEG-PCL nanoparticles. Biomaterials. 2012;**33**(32):8167-8176. DOI: 10.1016/j.biomaterials.2012.07.046

[62] Jiang X, et al. Solid tumor penetration by integrin-mediated pegylated poly(trimethylene carbonate) nanoparticles loaded with paclitaxel. Biomaterials. 2013;**34**(6):1739-1746. DOI: 10.1016/j.biomaterials.2012.11.016

[63] Huang K, et al. Size-dependent localization and penetration of ultrasmall gold nanoparticles in cancer cells, multicellular spheroids, and tumors in vivo. ACS Nano. 2012;**6**(5):4483-4493. DOI: 10.1021/nn301282m

[64] Jena PV, et al. Photoluminescent carbon nanotubes interrogate the permeability of multicellular tumor spheroids. Carbon N Y. 2016;**97**:99-109. DOI: 10.1016/j.carbon.2015.08.024

[65] Yong T, et al. Domino-like intercellular delivery of undecylenic acid-conjugated porous silicon nanoparticles for deep tumor penetration. ACS Applied Materials and Interfaces. 2016;**8**(41):27611-27621. DOI: 10.1021/acsami.6b11127

[66] Oliveira MS, et al. Solid lipid nanoparticles co-loaded with doxorubicin and α-tocopherol succinate are effective against drug-resistant cancer cells in monolayer and 3-D spheroid cancer cell models. International Journal of Pharmaceutics. 2016;**512**(1):292-300. DOI: 10.1016/j.ijpharm.2016.08.049

[67] Wang X, et al. Doxorubicin delivery to 3D multicellular spheroids and tumors based on boronic acid-rich chitosan nanoparticles. Biomaterials. 2013;**34**(19):4667-4679. DOI: 10.1016/j.biomaterials.2013.03.008

[68] Wang Y, et al. Tumor-penetrating nanoparticles for enhanced anticancer activity of combined photodynamic and hypoxia-activated therapy. ACS Nano. 2017;**11**(2):2227-2238. DOI: 10.1021/acsnano.6b08731

[69] Kulkarni P, et al. Hypoxia responsive, tumor penetrating lipid nanoparticles for delivery of chemotherapeutics to pancreatic cancer cell spheroids. Bioconjugate Chemistry. 2016;**27**(8):1830-1838. DOI: 10.1021/acs.bioconjchem.6b00241

[70] Thi Thuy Duong L, et al. Evaluation of anti-HER2 scFv-conjugated PLGA–PEG nanoparticles on 3D tumor spheroids of BT474 and HCT116 cancer cells. Advances in Natural Sciences: Nanoscience and Nanotechnology. 2016;**7**(2):025004

[71] Sarisozen C, et al. Nanomedicine based curcumin and doxorubicin combination treatment of glioblastoma with scFv-targeted micelles: In vitro evaluation on 2D and 3D tumor models. European Journal of Pharmaceutics and Biopharmaceutics. 2016;**108**:54-67. DOI: 10.1016/j.ejpb.2016.08.013

[72] Yao W, et al. Folic acid-conjugated soybean protein-based nanoparticles mediate efficient antitumor ability in vitro. Journal of Biomaterials Applications. 2016;**31**(6):832-843. DOI: 10.1177/0885328216679571

[73] Jiang Y, et al. Albumin-polymer conjugate nanoparticles and their interactions with prostate cancer cells in 2D and 3D culture: Comparison between PMMA and PCL. Journal of Materials Chemistry B. 2016;**4**(11):2017-2027. DOI: 10.1039/C5TB02576A

[74] Movia D, et al. Screening the cytotoxicity of single-walled carbon nanotubes using novel 3D tissue-mimetic models. ACS Nano. 2011;**5**(11):9278-9290. DOI: 10.1021/nn203659m

[75] Lee J, et al. In vitro toxicity testing of nanoparticles in 3D cell culture. Small. 2009;**5**(10): 1213-1221. DOI: 10.1002/smll.200801788

[76] Stocke NA, et al. Toxicity evaluation of magnetic hyperthermia induced by remote actuation of magnetic nanoparticles in 3D micrometastasic tumor tissue analogs for triple negative breast cancer. Biomaterials. 2017;**120**:115-125. DOI: 10.1016/j.biomaterials.2016.12.019

[77] Dias DR, Moreira AF, Correia IJ. The effect of the shape of gold core-mesoporous silica shell nanoparticles on the cellular behavior and tumor spheroid penetration. Journal of Materials Chemistry B. 2016;**4**(47):7630-7640. DOI: 10.1039/C6TB02668K

[78] Rane TD, Armani AM. Two-photon microscopy analysis of gold nanoparticle uptake in 3D cell spheroids. PLoS One. 2016;**11**(12):e0167548. DOI: 10.1371/journal.pone.0167548

[79] Zhao J, et al. Cellular uptake and movement in 2D and 3D multicellular breast cancer models of fructose-based cylindrical micelles that is dependent on the rod length. ACS Applied Materials and Interfaces. 2016;**8**(26):16622-16630. DOI: 10.1021/acsami.6b04805

[80] Bugno J, et al. Size and surface charge of engineered poly(amidoamine) dendrimers modulate tumor accumulation and penetration: A model study using multicellular tumor spheroids. Molecular Pharmaceutics. 2016;**13**(7):2155-2163. DOI: 10.1021/acs.molpharmaceut.5b00946

[81] Arranja A, et al. Interactions of Pluronic nanocarriers with 2D and 3D cell cultures: Effects of PEO block length and aggregation state. Journal of Controlled Release. 2016;**224**:126-135. DOI: 10.1016/j.jconrel.2016.01.014

[82] Li HJ, et al. Stimuli-responsive clustered nanoparticles for improved tumor penetration and therapeutic efficacy. Proceedings of the National Academy of Sciences of the United States of America. 2016;**113**(15):4164-4169. DOI: 10.1073/pnas.1522080113

[83] Corvaglia S, Guarnieri D, and Pompa PP. Boosting the therapeutic efficiency of nanovectors: Exocytosis engineering. Nanoscale. 2017;**9**(11):3757-3765. DOI: 10.1039/C7NR00364A

[84] Russell WMS, Burch RL. The Principles of Humane Experimental Technique. London, UK: Methuen & Co, Ltd; 1959

[85] Vinci M, et al. Advances in establishment and analysis of three-dimensional tumor spheroid-based functional assays for target validation and drug evaluation. BMC Biology. 2012;**10**:29. DOI: 10.1186/1741-7007-10-29

[86] Nyga A, Cheema U, Loizidou M. 3D tumour models: Novel in vitro approaches to cancer studies. Journal of Cell Communication and Signaling. 2011;**5**(3):239-248. DOI: 10.1007/s12079-011-0132-4

[87] Scannell JW, et al. Diagnosing the decline in pharmaceutical R&D efficiency. Nature Reviews Drug Discovery. 2012;**11**(3):191-200. DOI: 10.1038/nrd3681

[88] Mehta G, et al. Opportunities and challenges for use of tumor spheroids as models to test drug delivery and efficacy. Journal of Controlled Release. 2012;**164**(2):192-204. DOI: 10.1016/j.jconrel.2012.04.045

[89] Subramanian J, et al. Review of ongoing clinical trials in non-small-cell lung cancer: A status report for 2012 from the ClinicalTrials.gov Web site. Journal of Thoracic Oncology. 2013;**8**(7):860-865. DOI: 10.1097/JTO.0b013e318287c562

[90] Subramanian J, et al. Review of ongoing clinical trials in non-small cell lung cancer: A status report for 2009 from the ClinicalTrials.gov website. Journal of Thoracic Oncology. 2010;**5**(8):1116-1119. DOI: 10.1097/JTO.0b013e3181e76159

[91] Davies A. Cell Suspension Medium and Cell Suspension Medium Additive for the Three Dimensional Growth of Cells. 2013, Google Patents

[92] Zanoni M, et al. 3D tumor spheroid models for in vitro therapeutic screening: A systematic approach to enhance the biological relevance of data obtained. Scientific Reports. 2016;**6**:19103. DOI: 10.1038/srep19103

[93] Zschenker O, et al. Genome-wide gene expression analysis in cancer cells reveals 3D growth to affect ECM and processes associated with cell adhesion but not DNA repair. PLoS One. 2012;**7**(4):e34279. DOI: 10.1371/journal.pone.0034279

[94] Hirschhaeuser F, et al. Multicellular tumor spheroids: An underestimated tool is catching up again. Journal of Biotechnology. 2010;**148**(1):3-15. DOI: 10.1016/j.jbiotec.2010.01.012

[95] Costa EC, et al. 3D tumor spheroids: An overview on the tools and techniques used for their analysis. Biotechnology Advances. 2016;**34**(8):1427-1441. DOI: 10.1016/j.biotechadv.2016.11.002

[96] Rehfeldt F, et al. Cell responses to the mechanochemical microenvironment—Implications for regenerative medicine and drug delivery. Advanced Drug Delivery Reviews. 2007;**59**(13):1329-1339. DOI: 10.1016/j.addr.2007.08.007

[97] Elliott NT, Yuan F. A review of three-dimensional in vitro tissue models for drug discovery and transport studies. Journal of Pharmaceutical Sciences. 2011;**100**(1):59-74. DOI: 10.1002/jps.22257

[98] Liang Y, et al. A cell-instructive hydrogel to regulate malignancy of 3D tumor spheroids with matrix rigidity. Biomaterials. 2011;**32**(35):9308-9315. DOI: 10.1016/j.biomaterials.2011.08.045

[99] Shintani Y, et al. Collagen I promotes epithelial-to-mesenchymal transition in lung cancer cells via transforming growth factor-beta signaling. American Journal of Respiratory Cell and Molecular Biology. 2008;**38**(1):95-104. DOI: 10.1165/rcmb.2007-0071OC

[100] Kalluri R., Weinberg RA. The basics of epithelial-mesenchymal transition. Journal of Clinical Investigation. 2009;**119**(6):1420-1428. DOI: 10.1172/JCI39104

[101] Lamouille S, Xu J, Derynck R. Molecular mechanisms of epithelial-mesenchymal transition. Nature Reviews Molecular Cell Biology. 2014;**15**(3):178-196. DOI: 10.1038/nrm3758

[102] Richardson F, et al. The evaluation of E-Cadherin and vimentin as biomarkers of clinical outcomes among patients with non-small cell lung cancer treated with erlotinib as second- or third-line therapy. Anticancer Research. 2012;**32**(2):537-552

[103] Oliveira R, et al. Contribution of gap junctional communication between tumor cells and astroglia to the invasion of the brain parenchyma by human glioblastomas. BMC Cell Biology. 2005;**6**(1):7. DOI: 10.1186/1471-2121-6-7

[104] Osanai M, et al. Epigenetic silencing of occludin promotes tumorigenic and metastatic properties of cancer cells via modulations of unique sets of apoptosis-associated genes. Cancer Research. 2006;**66**(18):9125-9133. DOI: 10.1158/0008-5472.CAN-06-1864

[105] Jinn Y, Inase N. Connexin 43, E-cadherin, beta-catenin and ZO-1 expression, and aberrant methylation of the connexin 43 gene in NSCLC. Anticancer Research. 2010;**30**(6): 2271-2278

[106] Hazrin-Chong NH, Manefield M. An alternative SEM drying method using hexamethyldisilazane (HMDS) for microbial cell attachment studies on sub-bituminous coal. Journal of Microbiological Methods. 2012;**90**(2):96-99. DOI: 10.1016/j.mimet.2012.04.014

[107] Braet F, De Zanger R, Wisse E. Drying cells for SEM, AFM and TEM by hexamethyldisilazane: A study on hepatic endothelial cells. Journal of Microscopy. 1997;**186**(Pt 1):84-87

[108] Bell DC. Contrast mechanisms and image formation in helium ion microscopy. Microscopy and Microanalysis. 2009;**15**(2):147-153. DOI: 10.1017/S1431927609090138

[109] Bazou D, et al. Elucidation of flow-mediated tumour cell-induced platelet aggregation using an ultrasound standing wave trap. British Journal of Pharmacology. 2011;**162**(7):1577-1589. DOI: 10.1111/j.1476-5381.2010.01182.x

[110] Rice WL, et al. High resolution helium ion scanning microscopy of the rat kidney. PLoS One. 2013;**8**(3):e57051. DOI: 10.1371/journal.pone.0057051

[111] Bazou D, et al. Imaging of human colon cancer cells using He-Ion scanning microscopy. Journal of Microscopy. 2011;**242**(3):290-294. DOI: 10.1111/j.1365-2818.2010.03467.x

[112] Mohamed BM, et al. Citrullination of proteins: A common post-translational modification pathway induced by different nanoparticles in vitro and in vivo. Nanomedicine (London, England). 2012;**7**(8):1181-1195. DOI: 10.2217/nnm.11.177

[113] Bauer H, et al. The dual role of zonula occludens (ZO) proteins. Journal of Biomedicine & Biotechnology. 2010;**2010**:402593. DOI: 10.1155/2010/402593

[114] Friedrich J, et al. Spheroid-based drug screen: Considerations and practical approach. Nature Protocols. 2009;**4**(3):309-324. DOI: 10.1038/nprot.2008.226

[115] Ju RJ, Mu LM, Lu WL. Targeting drug delivery systems for circumventing multidrug resistance of cancers. Therapeutic Delivery. 2013;**4**(6):667-671. DOI: 10.4155/tde.13.39

[116] Kim H, Phung Y, Ho M. Changes in global gene expression associated with 3D structure of tumors: An ex vivo matrix-free mesothelioma spheroid model. PLoS One. 2012;**7**(6):e39556. DOI: 10.1371/journal.pone.0039556

[117] Movia D, et al. A safe-by-design approach to the development of gold nanoboxes as carriers for internalization into cancer cells. Biomaterials. 2014;**35**(9):2543-2557. DOI: 10.1016/j.biomaterials.2013.12.057

[118] Ivascu A, Kubbies M. Rapid generation of single-tumor spheroids for high-throughput cell function and toxicity analysis. Journal of Biomolecular Screening. 2006;**11**(8):922-932. DOI: 10.1177/1087057106292763

[119] Del Duca D, Werbowetski T, Del Maestro RF. Spheroid preparation from hanging drops: Characterization of a model of brain tumor invasion. Journal of Neuro-Oncology. 2004;**67**(3):295-303

[120] Neto AI, et al. A novel hanging spherical drop system for the generation of cellular spheroids and high throughput combinatorial drug screening. Biomaterials Science. 2015;**3**(4):581-585. DOI: 10.1039/c4bm00411f

[121] Ivascu A, Kubbies M. Diversity of cell-mediated adhesions in breast cancer spheroids. International Journal of Oncology. 2007;**31**(6):1403-1413

[122] Xiang X, et al. The development and characterization of a human mesothelioma in vitro 3D model to investigate immunotoxin therapy. PLoS One. 2011;**6**(1):e14640. DOI: 10.1371/journal.pone.0014640

[123] Wright MH, et al. Brca1 breast tumors contain distinct CD44+/CD24- and CD133+ cells with cancer stem cell characteristics. Breast Cancer Research. 2008;**10**(1):R10. DOI: 10.1186/bcr1855

[124] Takaishi S, et al. Identification of gastric cancer stem cells using the cell surface marker CD44. Stem Cells. 2009;**27**(5):1006-1020. DOI: 10.1002/stem.30

[125] Cheng V, et al. High-content analysis of tumour cell invasion in three-dimensional spheroid assays. Oncoscience. 2015;**2**(6):596-606. DOI: 10.18632/oncoscience.171

[126] Matsuda Y, et al. Morphological and cytoskeletal changes of pancreatic cancer cells in three-dimensional spheroidal culture. Medical Molecular Morphology. 2010;**43**(4):211-217. DOI: 10.1007/s00795-010-0497-0

[127] Hardelauf H, et al. Microarrays for the scalable production of metabolically relevant tumour spheroids: A tool for modulating chemosensitivity traits. Lab on a Chip. 2011;**11**(3):419-428. DOI: 10.1039/c0lc00089b

[128] Fischbach C, et al. Engineering tumors with 3D scaffolds. Nature Methods. 2007;**4**(10): 855-860. DOI: 10.1038/nmeth1085

[129] Wartenberg M, et al. Tumor-induced angiogenesis studied in confrontation cultures of multicellular tumor spheroids and embryoid bodies grown from pluripotent embryonic stem cells. FASEB Journal. 2001;**15**(6):995-1005

[130] Unsworth BR, Lelkes PI. Growing tissues in microgravity. Nature Medicine. 1998;**4**(8): 901-907

[131] Mazzoleni G, Di Lorenzo D, Steimberg N. Modelling tissues in 3D: The next future of pharmaco-toxicology and food research? Genes & Nutrition. 2009;**4**(1):13-22. DOI: 10.1007/s12263-008-0107-0

[132] Thoma CR, et al. A high-throughput-compatible 3D microtissue co-culture system for phenotypic RNAi screening applications. Journal of Biomolecular Screening. 2013;**18**(10):1330-1337. DOI: 10.1177/1087057113499071

[133] HTStec. 3D Cell Culture Trends 2011. 2011. Available at: http://www.htstec.com/consultancyitem.aspx?Item=374

[134] Graf BW, Boppart SA. Imaging and analysis of three-dimensional cell culture models. Methods in Molecular Biology. 2010;**591**:211-227. DOI: 10.1007/978-1-60761-404-3_13

[135] http://www.perkinelmer.com/pdfs/downloads/APP_Opera-Microtissue-Cores.pdf [Accessed on: March 2017]

[136] Hama H, et al. Scale: A chemical approach for fluorescence imaging and reconstruction of transparent mouse brain. Nature Neuroscience. 2011;**14**(11):1481-1488. DOI: 10.1038/nn.2928

[137] Zhang JZ, et al. The use of spectroscopic imaging and mapping techniques in the characterisation and study of DLD-1 cell spheroid tumour models. Integrative Biology (Camb). 2012;**4**(9):1072-1080. DOI: 10.1039/c2ib20121f

[138] https://3dbiomatrix.com/wp-content/uploads/2013/10/Perfecta3D-HDP1096-Protocols_SF_Spheroid-Formation_05_13.pdf [Accessed on: January 2017]

[139] https://3dbiomatrix.com/wp-content/uploads/2013/10/Perfecta3D-HDP1096-Protocols_SF_Exchange_12-15-12.pdf [Accessed on: January 2017]

[140] https://3dbiomatrix.com/wp-content/uploads/2014/02/Perfecta3D-Protocols_SF_Transfer-96well-2_2014.pdf [Accessed on: January 2017]

[141] Rothen-Rutishauser B, et al. In vitro models of the human epithelial airway barrier to study the toxic potential of particulate matter. Expert Opinion on Drug Metabolism & Toxicology. 2008;**4**(8):1075-1089. DOI: 10.1517/17425255.4.8.1075

[142] Caino MC, et al. Non-small cell lung carcinoma cell motility, rac activation and metastatic dissemination are mediated by protein kinase C epsilon. PLoS One. 2012;**7**(2):e31714. DOI: 10.1371/journal.pone.0031714

[143] Gazdar AF, et al. Lung cancer cell lines as tools for biomedical discovery and research. Journal of the National Cancer Institute. 2010;**102**(17):1310-1321. DOI: 10.1093/jnci/djq279

[144] Mohamed BM, et al. Citrullination as early-stage indicator of cell response to single-walled carbon nanotubes. Scientific Reports. 2013;**3**:1124. DOI: 10.1038/srep01124

[145] Mohamed BM, et al. Activation of stress-related signalling pathway in human cells upon SiO_2 nanoparticles exposure as an early indicator of cytotoxicity. Journal of Nanobiotechnology. 2011;**9**:29. DOI: 10.1186/1477-3155-9-29

[146] Verma NK, et al. Autophagy induction by silver nanowires: A new aspect in the biocompatibility assessment of nanocomposite thin films. Toxicology and Applied Pharmacology. 2012;**264**(3):451-461. DOI: 10.1016/j.taap.2012.08.023

[147] Verma NK, et al. Magnetic core-shell nanoparticles for drug delivery by nebulization. J Nanobiotechnology. 2013;**11**:1. DOI: 10.1186/1477-3155-11-1

[148] Singh M, et al. Silver nanowires as prospective carriers for drug delivery in cancer treatment: An in vitro biocompatibility study on lung adenocarcinoma cells and fibroblasts. European Journal of Nanomedicine. 2013;**5**(4):195-204. DOI: 10.1515/ejnm-2013-0024

[149] Stratmann AT, et al. Establishment of a human 3D lung cancer model based on a biological tissue matrix combined with a Boolean in silico model. Molecular Oncology. 2014;**8**(2):351-365. DOI: 10.1016/j.molonc.2013.11.009

[150] Godugu C, et al. AlgiMatrix based 3D cell culture system as an in-vitro tumor model for anticancer studies. PLoS One. 2013;**8**(1):e53708. DOI: 10.1371/journal.pone.0053708

[151] Cichon MA, et al. Growth of lung cancer cells in three-dimensional microenvironments reveals key features of tumor malignancy. Integrative Biology (Camb). 2012;**4**(4):440-448. DOI: 10.1039/c1ib00090j

[152] Xu F, et al. A three-dimensional in vitro ovarian cancer coculture model using a high-throughput cell patterning platform. Biotechnology Journal. 2011;**6**(2):204-212. DOI: 10.1002/biot.201000340

[153] Friedrich J, Ebner R, Kunz-Schughart LA. Experimental anti-tumor therapy in 3-D: Spheroids--old hat or new challenge? International Journal of Radiation Biology. 2007;**83**(11-12):849-871. DOI: 10.1080/09553000701727531

[154] Nakamura K, et al. Apoptosis induction of human lung cancer cell line in multicellular heterospheroids with humanized antiganglioside GM2 monoclonal antibody. Cancer Research. 1999;**59**(20):5323-5330

[155] Krueger S, et al. Interactions between human colon carcinoma cells, fibroblasts and monocytic cells in coculture—Regulation of cathepsin B expression and invasiveness. Cancer Letters. 2005;**223**(2):313-322. DOI: 10.1016/j.canlet.2004.09.050

[156] Paduch R and Niedziela P. TGF-beta1 influence on TNF-alpha production and sTNF-Rs shedding in a coculture of colon carcinoma cell spheroids with normal cells. In Vitro Cellular and Developmental Biology—Animal. 2009;**45**(7):371-377. DOI: 10.1007/s11626-009-9190-9

[157] Brouty-Boye D, et al. Fibroblast-mediated differentiation in human breast carcinoma cells (MCF-7) grown as nodules in vitro. International Journal of Cancer. 1994;**56**(5):731-735

[158] Seidl P, et al. Three-dimensional fibroblast-tumor cell interaction causes downregulation of RACK1 mRNA expression in breast cancer cells in vitro. International Journal of Cancer. 2002;**102**(2):129-136. DOI: 10.1002/ijc.10675

[159] Schuster U, et al. A heterologous in vitro coculture system to study interaction between human bladder cancer cells and fibroblasts. Journal of Urology. 1994;**151**(6):1707-1711

[160] Kunz-Schughart LA, et al. A heterologous 3-D coculture model of breast tumor cells and fibroblasts to study tumor-associated fibroblast differentiation. Experimental Cell Research. 2001;**266**(1):74-86. DOI: 10.1006/excr.2001.5210

[161] Kim SH, et al. Human lung cancer-associated fibroblasts enhance motility of non-small cell lung cancer cells in co-culture. Anticancer Research. 2013;**33**(5):2001-2009

[162] An J, et al. Significance of cancer-associated fibroblasts in the regulation of gene expression in the leading cells of invasive lung cancer. Journal of Cancer Research and Clinical Oncology. 2013;**139**(3):379-388. DOI: 10.1007/s00432-012-1328-6

[163] Walzl A, et al. The resazurin reduction assay can distinguish cytotoxic from cytostatic compounds in spheroid screening assays. Journal of Biomolecular Screening. 2014;**19**(7):1047-1059. DOI: 10.1177/1087057114532352

[164] Ng KW, Leong DT, Hutmacher DW. The challenge to measure cell proliferation in two and three dimensions. Tissue Engineering. 2005;**11**(1-2):182-191. DOI: 10.1089/ten.2005.11.182

[165] Movia D, Giordani S. Toxicity of carbon nanotubes. In: Handbook of Green Chemistry. Wiley-VCH Verlag GmbH & Co. KGaA; Weinheim, Germany. 2010. DOI: 10.1002/9783527628698.hgc091

[166] Cho EC, Zhang Q, Xia Y. The effect of sedimentation and diffusion on cellular uptake of gold nanoparticles. Nature Nanotechnology. 2011;**6**(6):385-391. DOI: 10.1038/nnano.2011.58

Chapter 6

Applications of Fluorescent Quantum Dots for Reproductive Medicine and Disease Detection

Sapna Jain, Seong B. Park, Shreekmar R. Pillai,
Peter L. Ryan, Scott T. Willard and Jean M. Feugang

Additional information is available at the end of the chapter

http://dx.doi.org/10.5772/intechopen.72978

Abstract

Understanding the mechanisms associated with fertility and disease management in animals remains challenging. Continuing advances in nanotechnology provide new tools and alternative approaches for the investigation of these mechanisms. Fluorescent quantum dot nanoparticles, for example, have unique physicochemical properties, which allow for *in vivo* and *in vitro* imaging in various areas of life sciences. Traditional quantum dots contain heavy metal semiconductor cores, which have raised concern over their potential for toxicity. The majority of available quantum dots today prevent heavy metal release with additional chemical and polymer layers for safe water solubility. In this chapter, the most widely used quantum dots made of cadmium selenide, which possess great potential for real-time imaging in disease detection and reproductive medicine, are discussed.

Keywords: quantum dots, spermatozoa, *in vivo* imaging, real-time imaging, luminescence, fertility

1. Introduction

Since their discovery by Alexie Ekimov and Louis Brus in the 1980s, quantum dot (QD) nanoparticles have been categorized as a novel class of fluorescent particles [1]. Fluorescent nanoparticles exhibit distinct energy levels and size-dependent fluorescent emission [2]. The QD sizes range from 2 to 10 nm (10–50 atoms) in diameter, with the smaller size corresponding to the larger bandgap [3]. Each QD absorbs white light and then reemits a specific color associated with the material's bandgap, from blue to red or near-infrared (NIR) as the QD crystals increase in size [4]. The variety of fluorescence emission is very useful for both *in vitro* and *in vivo* multiplex bioimaging as multiple QDs can be used in one subject or field of view to image a variety of targets under a single excitation.

QDs have unique advantages over traditional dyes and fluorescent proteins such as a high quantum yield, extreme brightness, tunable emission wavelength, long fluorescence duration, exceptional photostability and resistance to photobleaching [5]. In addition, their high extinction coefficient makes them ideal for optical applications and transport. Since QDs wavelengths are tunable based on size, their conducting properties can be very well controlled to suit various applications. Zinc sulfide (ZnS)-coated cadmium selenide (CdSe) nanocrystals are the most commonly studied QDs for bioapplications due to their wide bandgap and easily tunable emission in the visible range [6]. These qualities make them especially useful for various industrial, agricultural, and biomedical applications [7]. Another reason for the popularity of the CdSe QDs is their well-established synthesis and characterization protocols [8]. In this chapter, the synthesis, toxicity, and surface modification of CdSe QDs in bioanalytics and biomedical diagnostics are discussed.

1.1. Synthesis of QDs

Quantum dots can be prepared by formation of nanosized semiconductor particles through colloidal chemistry or by epitaxial growth and/or nanoscale patterning [9]. Preparation of QDs designed for biological applications has four basic steps: core synthesis, shell growth, aqueous solubilization, and biomolecular conjugation or biofunctionalization.

1.1.1. Core-shell protocol

QDs core is generally made from heavy metal semiconductors of group II–VI (CdSe, CdS, CdTe, HgS, ZnS, ZnSe), III–V (GaAs, GaN, InP, InAs, InGaAs), IV–VI (PbS, PbSe, PbTe, SnTe), and group III–V (InP and InGaP) (**Table 1**). The most common method for preparation of QDs core consists of a rapid injection of semiconductor or organometallic precursors (e.g., Cd precursor and TOPSe) into hot and vigorously stirred specific coordinating solvent (e.g., thiol stabilizers). Coordinating solvents stabilize the bulk semiconductors and avoid aggregation as the QDs grow [10]. Thereafter, the semiconductor core material (e.g., CdSe) must be protected from degradation and oxidation to optimize QDs performance. Hence, an external layer or protective shell (e.g., ZnS) is usually synthesized to cover the QD semiconductor core to enhance stability, while increasing its photoluminescence [11]. Due to their synthesis in nonpolar organic solvents, the inorganic core-shell semiconductor QDs (e.g., CdSe) are typically hydrophobic, which prevents their solubility and enhances the formation of aggregates or precipitates in water-based solutions. This property limits biological applications of core-shell QDs, requiring additional modifications of their surfaces to achieve biocompatibility or solubility in biological or water-based fluids.

1.1.2. Aqueous solubilization

The aqueous dispersal of core-shell QDs is controlled by the chemical nature of their surface coating. Numerous effective methods have been established to create hydrophilic QDs, which can be divided into two main categories [12]. *The first route*, commonly designated as

http://dx.doi.org/10.5772/intechopen.72978

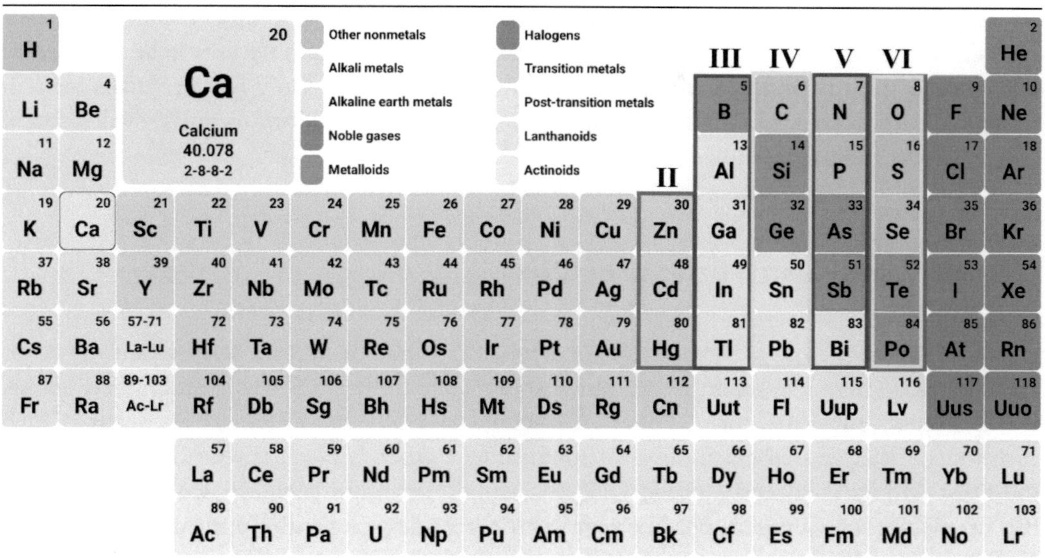

Table 1. Semiconductor elements—groups II to VI—within the periodic table.

"cap exchange procedure," consists of a complete replacement of the hydrophobic layer of organic solvent by bioactive molecules containing soft acidic and hydrophilic groups pointing outwards, from the QDs surface to surrounding bulk water molecules [13, 14]. This route allows electrostatic stabilization of inorganic core shell of QDs through their interactions with small charged ligands (e.g., amines, cystamine, cysteine, 2-mercaptoethanol, ethylamine, or mercaptopropionate) or charged surfactants to form a new external coating layer that encapsulates QDs. *The second route* allows steric stabilization through modification of the native coordinating organic ligands on the QDs surface with "bulky" uncharged polymeric surface ligands such as the polyethylene glycol or PEG [15, 16]. *Alternatively to electrostatic and steric stabilizations,* bulky and charged ligands (e.g., polyelectrolytes or polyethyleneimine), amphiphilic inorganic shell (e.g., silica added to QDs during polycondensation) or solid lipid nanoparticles composed of high biocompatible lipids of physical and chemical long-term stability have been successfully tested for further stabilization of QDs [17–19]. All aforementioned coating strategies are useful for QDs solubilization while allowing further addition of polymers or bioactive molecules for cell labeling and imaging.

1.1.3. Biofunctionalization

Biofunctionalization refers to the ability to successfully attach or conjugate bioactive molecules (e.g., oligonucleotides, proteins, polysaccharide, and peptides) to water-dispersed QDs. This process can be achieved by binding to polyhistidine tags, electrostatic (e.g., avidin-biotin) or covalent interactions. This later is typically accomplished by activated 1-ethyl-3-(3-dimethylaminopropyl) carbodiimide (EDC) coupling amine and carboxyl groups and catalyzed maleimide (SMCC) linking amine to sulfhydryl groups [20, 21]. It is important to mention that these processes remain challenging due to surface chemistry of QDs, control of attachment orientation of biomolecules [19, 21, 22], and determination of conjugation efficacy. QDs have greater surface area-to-volume ratio allowing several types of biomolecules to be attached to a single QD to provide multifunctionality of the conjugate [23]. Only few protocols for biofunctionalization are available, and systematic studies are needed for functional evaluation of conjugated or biofunctionalized QDs [14, 20, 21].

2. *In vitro* and *in vivo* toxicity of QDs

Due to their heavy metal semiconductor cores, QDs are considered toxic when the cores are not adequately contained by an outer shell, such as the ZnS shells mentioned above. Without containing the cores, potential damage to biological systems can occur, which composes a challenge to surmount for medical and other *in vivo* applications. The core of the most widely used and studied QDs consists of cadmium selenide (CdSe) or telluride (CdTe) given their quantum confinement region spanning the entire optical spectrum [24]. Cadmium ions (Cd^{2+}) have been identified as the primary cause of QDs cytotoxicity due to their overtime leaking, upon illumination or oxidation [25, 26]. Leaked Cd^{2+} is able to bind to thiol groups of key molecules of mitochondria and cause enough stress and damage leading to cell death [27].

Moreover, the cytotoxicity of QDs appears directly related to the protective inorganic surface layers [25]. Additional surface coatings may be needed to substantially reduce or eliminate the release of Cd^{2+} [26]. The utilization of gelatin during the production of CdTe QDs has resulted in reduced toxicity of particles [28]. In the case of CdSe QDs, it is believed that properly prepared closed (ZnS) or multiple (e.g., ZnS/SiO_2, ZnS/PEG hydrophilic coating) shells render cadmium leakage less likely [29, 30]. However, oxidized QDs surface may unintendedly react with intracellular components, causing formation and release of reduced Cd that results in apoptosis within primary hepatocytes isolated from rats [31–33]. In addition, the charge and chemical reactivity of QDs play dominant roles in their biocompatibility, independent of their size [32]. Various studies have demonstrated the crucial roles of multiple positive charges and size-dependent polycationic materials of QDs in cytotoxic mechanisms of nanoparticles [34, 35].

Majority of *in vitro* studies use transformed cell lines to demonstrate the cytotoxicity of QDs that may not fully reflect the response cascade in normal cells [12]. Nonetheless, the use of these cell types allows for many generalizations to be made regarding the toxicity related to specific QDs features (i.e., size, protective shell, and surface chemistry), experimental dosage, and exposure conditions. **Table 2** summarizes few studies exemplifying the complexity of investigating QDs nanotoxicity due to multiple variables such as their size, shell components and surface chemistry that should be taken into account when designing an experiment. The comparable size of QDs with certain cellular components may facilitate their passage through many biological barriers and accumulation in different tissues to cause adverse effects after long-term exposure [36]. At equal concentrations and positive charges, smaller QDs (i.e., 2–3 nm) display high cytotoxicity than larger ones (i.e., >5 nm), with liver and kidneys often being main target organs due to their blood filtering function [37].

QDs (core and protective layers)	Concentration	Exposure	Toxic effect	References
CdSe/ZnS-SSA	0.1–0.4 mg/mL	0–24 h	0.1 mg/mL altered cell growth; most cells nonviable at 0.4 mg/mL	[38]
CdSe/ZnS-SSA	0.1 mg/mL QDs per 5×10^7 cells	2 h to 7 days	No toxicity in mice *in vivo*	[38]
CdSe/ZnS conjugates: NH_2, OH, OH/COOH, NH_2/OH, COOH	1–2 μM	12 h	2-μM QD-COOH-induced DNA damage upon 2 h of exposure	[39]
CdSe/ZnS/MUA	0–0.4 mg/mL	24 h	0.2 mg/mL, Vero; 0.1 mg/mL, HeLa; 0.1 mg/mL, hepatocytes	[40]
CdTe	0.01–100 μg/mL	2–24 h	10 μg/mL cytotoxic	[41]
CdSe-MAA, TOPO QDs	62.5–1000 μg/mL	1–8 h	62.5 μg/mL cytotoxic under oxidative/photolytic conditions	[25]
			No toxicity on addition of ZnS cap	
QD micelles: CdSe/ZnS QDs in (PEG-PE) and phophatidylcholine	1.5–3 nL of 2.3-μM QDs injected, approx. 2.1×10^9 to 4.2×10^9 QDs/cell	Days	5×10^9 QDs/cell: cell abnormalities, altered viability and motility	[42]
			No toxicity at 2×10^9 QDs/cells	
CdSe/ZnS amp-QDs and mPEG QDs	Injections, approx. 180-nm QD, approx. 20-pmol QD/g animal weight	15-min cells incubation, 1–133 days *in vivo*	No signs of localized necrosis at the sites of deposition	[43]
CdSe/ZnS-DHLA	400–600 nM	45–60 min	No effect on cell growth	[44]
Avidin-conjugated CdSe/ZnS QDs	0.5–1.0 μM	15 min	No effect on cell growth and development	[44]
CdSe/ZnS-amphiphilic micelle	60-μM QD/g animal weight, 1-μM and 20-nM final QD concentration	Information not provided	Mice showed no noticeable ill effects after imaging	[45]
CdSe/ZnS-DHLA QDs	100 μL of B16F10 cells (approx. 2×10^5 to 4×10^5) used for tail vein injection	4–6 h cell incubation, mice sacrificed at 1–6 h	No toxicity observed in cells or mice	[46]
CdSe/ZnS-MUA QDs; QD-SSA complexes	0.24 mg/mL	2 h	0.4 mg/mL MUA/SSA-QD complexes did not affect viability Vero cells	[47]
CdSe/ZnS	10-pmol QDs/1×10^5 cells (approx. 10 nM)	10 days (cell culture)	10 nM QD had minimal impact on cell survival	[48]
CdTe aqQDs	300–600 nM	3 days	Nearly completely inhibited cell growth even from the very beginning	[49]
CdTe-gelatinized/ nongelatinized	1–100 nM	72 h	At 1 nM, did not initiate any detrimental effects; at 100 nM, resulted in the death of all PC12 cells	[50]

QDs (core and protective layers)	Concentration	Exposure	Toxic effect	References
CdTe, CdTe/CdS, CdTe/CdS/ZnS	0.2–0.3 μM	0–48 h	Cells treated with CdTe and CdTe/CdS QDs were mostly nonviable by 48 h (for all concentrations tested)	[51]
CdSe/ZnS-PEG (EviTag T1 490 QD)	0.84–105 μM	0–24 h	Commercially available QDs demonstrated low cytotoxicity but induced cell detachment	[52]
CdSe	1, 10, and 20 nM	24 h	1 nM QD for 24 h showed no decreased in cell viability; in contrast, cells treated with 10 and 20 nM QDs for 24 h showed decreases in cell viability in the order of 20 and 30%	[53]

Table 2. Inconsistent considerations on QDs toxicity evaluation (modified from [3]).

Due to the complexity to characterize the cytotoxicity of nanoparticles, the US National Cancer Institute and several other US health agencies have created the Nanotechnology Characterization Laboratory (NCL) for efficacy and toxicity testing of nanoparticles, including fluorescent QDs. As part of the process, the NCL will describe physical attributes of nanoparticles, their *in vitro* biological properties and their *in vivo* biocompatibility.

3. Bioapplications

Despite the reported and controversial cytotoxicity, QD nanoparticles remain excellent candidates for numerous bioapplications. Compared to organic dyes, QDs display narrow, symmetrical, and tunable emission spectra and contingency for their size and material composition [54]. Various QD sizes have closer but nonoverlapping emission wavelengths [20], which excitation through a single light source leads to a photostable and broad absorption spectra [52].

3.1. QDs labeling

The brightness of CdSe QDs fluorescence has made them the widely labeled nanoparticles for various biosensing (e.g., oligonucleotides, organic dyes) and single or multiplex labeling (e.g., antibodies, peptides) [22, 23, 31, 55–57]. Yet, the localization (intracellular or extracellular), expression level, and environment (oxidizing or reducing) of the target molecule should be considered during the QDs labeling protocol. For example, the intracellular targeting may pose additional challenges requiring the need of cell-penetrating peptides (e.g., polyarginine, polylysine) for effective intracellular delivery of QDs conjugates, while maintaining the homeostasis and osmotic balance of cells. Reproductive studies have shown the ability of porcine gametes to interact with self-illuminated CdSe/ZnS QDs [58, 59] with the necessity to determine the suitable sperm-to-QDs ratio, avoiding or limiting QDs toxicity to sperm function, as observed in previous studies using various nanoparticles [58–65]. *In vitro* matured oocytes appeared to accumulate higher levels of QDs compared to spermatozoa, which, instead, exhibited stronger membrane labeling (**Figures 1** and **2**). Reduced QDs internalization within the spermatozoa was attributed to sperm membrane specificities, whereas the limitations of QDs as compared to organic dyes may not be ruled out [66]. The conjugation of self-illuminated CdSe/ZnS QDs with anti-plasminogen antibody (for specific targeting) revealed stronger signals within the porcine oocyte than the nonconjugated QDs, applied for plain targeting (**Figure 2**). The use of these CdSe/ZnS QDs also provided opportunity for *ex vivo* imaging of cultured porcine ovarian follicles (**Figure 3**), which would, in the near future, permit real-time monitoring of key molecules having role(s) during folliculogenesis.

3.1.1. QDs labeling for cell imaging and disease detection

Effective labeling of fluorescent QDs is crucial for extracellular and intracellular tracking of target molecules in their native environment. QDs functionalized with antibody are optimal for extracellular targeting of cell-surface membrane proteins (e.g., receptors) and subsequent targeted imaging [20, 22, 31], which practice will create opportunities for precise assessments of cellular and molecular mechanisms of diseases (e.g., cancer) and their treatments. Near-infrared QDs (e.g., CdSe, CdTe) emit in the wavelength range of 650–900 nm to overcome the optical property variations and endogenous autofluorescence of tissues under *in vivo* conditions [67], permitting tumor localization and visualization while offering a new mean for cancer prevention and treatment.

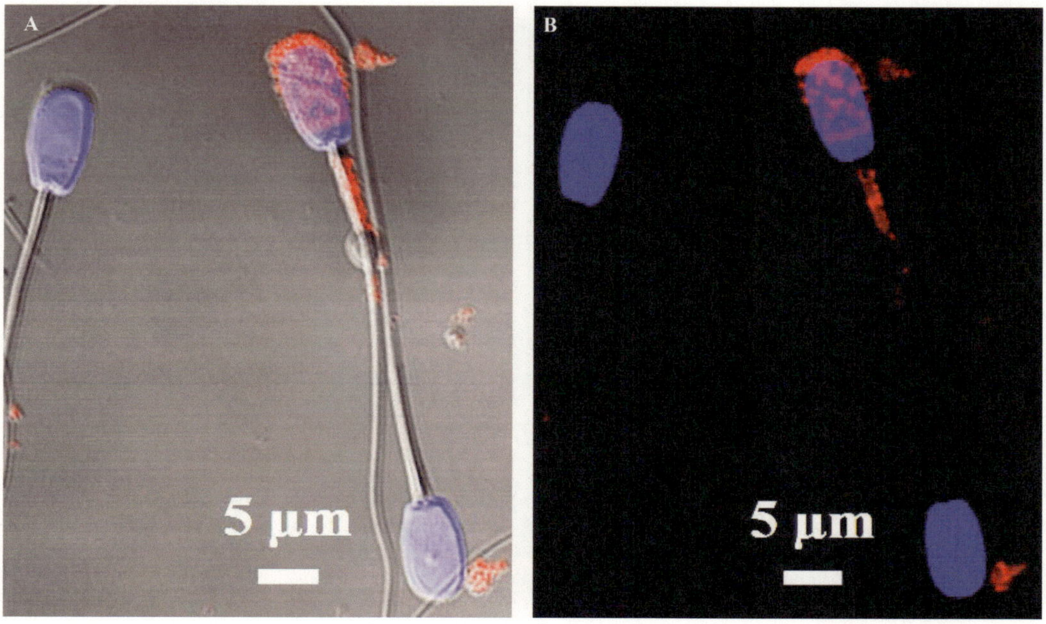

Figure 1. Confocal microscope imaging of mature boar spermatozoa labeled with CdSe QDs 655 nm. Labeled spermatozoa revealed major localizations of QDs (red spots) in the head and mid-piece regions. Sperm nuclei are counterstained in blue with DAPI. Micrograph A = overlay of 3 lights (visible, blue DAPI and red QDs 655 nm); Micrograph B = overlay of DAPI and QD 655 nm.

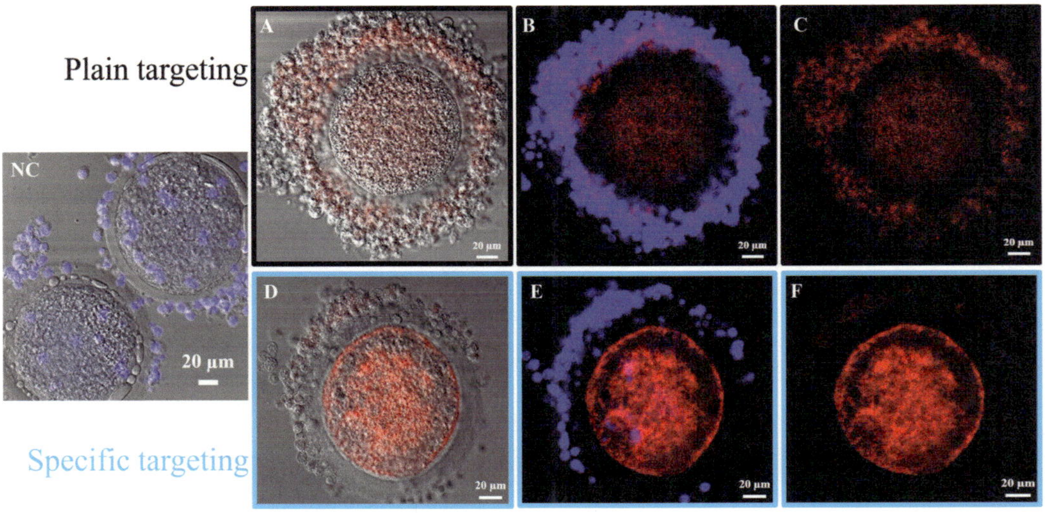

Figure 2. Confocal microscope imaging of porcine oocytes matured in the presence of QDs 655 nm. Cumulus-oocyte complexes were matured in the presence of QD alone (plain targeting; micrographs A/B/C) or QD conjugated with anti-plasminogen antibody (specific targeting: micrographs D/E/F). Micrograph NC = Control without QDs; A and D = Overlays of visible light and QDs 655 nm filter; B and D = overlays of DAPI and QDs 655 nm, and C and F = QDs 655 nm filter alone. The stronger and differential distribution of the red signal can be seen following QDs conjugation.

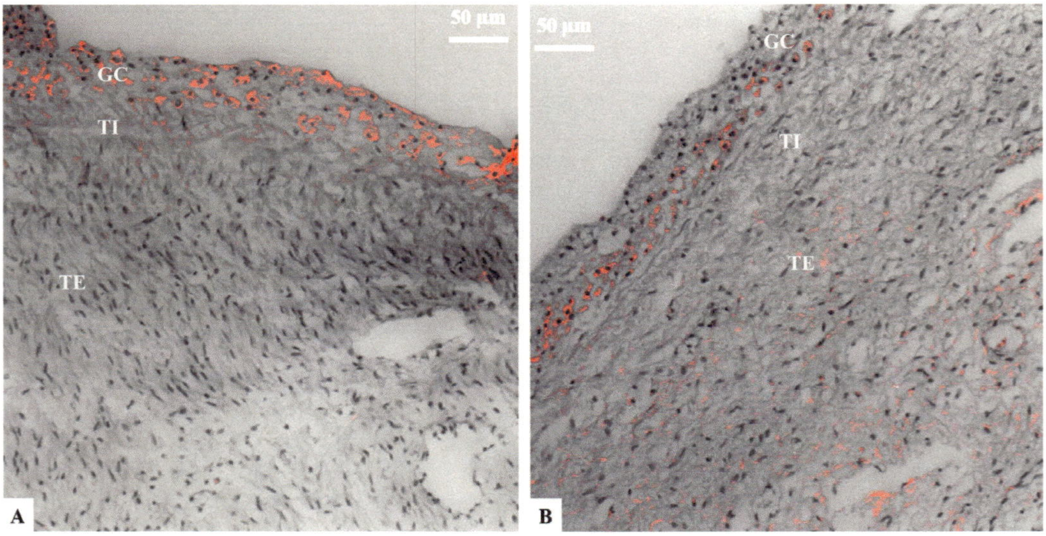

Figure 3. Confocal microscope imaging of porcine follicles microinjected with nonconjugated QDs 655 nm. Dissected antral follicles were microinjected, cultured for 1 (A) or 3 (B) days, then prepared for histology slides and imaging. QDs (red sports) are mainly visible within the layer of granulosa cells (GC) after 1 day of culture (A) and then throughout the theca interna (TI) and externa (TE) after 3 days of culture (B).

The intracellular localization of selective biomolecules for targeting presents additional challenges associated with QDs conjugates delivery within the cells. Available methods for QDs delivery are composed of, but not limited to, positively charged peptides or cell-penetrating peptides on QDs, microinjection, electroporation, or nonspecific or receptor-mediated endocytosis

[57, 67–69]. Electroporation technique has shown robust and highly efficient delivery of both monomer and aggregate QDs to the cells due to induced electrical pulses that temporarily permeabilize the plasma membrane [70]. It has been used for *in vivo* imaging of cancer cells through active intracellular delivery of QDs [67]. Electroporation of QDs in lung (NCI-H460) and ovary (SK-OV-3) cancer cells revealed high and longer (over a month) QDs retention inside the cells, allowing observation of the entire process of subcutaneous tumor growth and cancer cell dissemination at late stages of metastasis in a natural tissue environment [67]. It is important to mention that biofunctionalized QDs have been used for imaging in many other diseases, including the brain tissue [71, 72].

3.1.2. QDs labeling for cell imaging in reproductive biology

The small size (2–10 nm in diameter) and unique physicochemical properties of QDs, especially their tunable size-dependent fluorescence emission, make them excellent candidates for applications in the reproductive field. For example, the multicolor detection of various QDs permits spectral multiplexing for simultaneous detection and quantification of different biomolecules in *in vitro* bioassays [23, 54, 73–75], which may be crucial in understanding the complexity of mammalian gamete maturation in their native environment.

Additionally, signal amplification of enzymatic reactions could be achieved through QDs emitting localized and bright fluorescence in bioassays. This later could be illustrated by the novel QD-BRET (Bioluminescent Resonance Energy Transfer), a luciferase-doped QDs which enzymatic reaction with its substrate (luciferin or coelenterazine) produces energy (480 nm) that is immediately absorbed by the CdSe QDs to emit brighter and long-lasting NIR fluorescence [76]. Numerous *in vivo, in vitro,* and *ex situ* studies have successfully applied the QD-BRET for imaging of somatic and reproductive (mammalian gametes and ovarian follicle; **Figures 1–3**) cells [58, 59, 77].

The use of spectral multiplexing and signal amplification in reproductive biology has potential for fast diagnostics of gamete quality through direct (e.g., fluoroimmunoassays) or indirect (lab-on-chip arrays) evaluations. The proposed arrays should contain various QDs sizes that are biofunctionalized to target various key biomarkers of reproductive cells.

In addition to above-mentioned applications, there is potential to use QDs conjugates for targeted labeling, tracking, and imaging of ovarian follicle cells (during folliculogenesis) or spermatozoa (during intrauterine migration). Moreover, a recent study using amphibians reported the ability of living tadpoles to accumulate QD (655 nm) nanoparticles, likely impeding with their development [78].

4. Future outlook

QDs applications have been explored for molecular and pharmaceutical fields, but are rapidly expanding to other research areas. It is expected that QDs will be used for (1) categorizing various types of biological processes, (2) localizing and identifying molecular mechanisms of disease, (3) developing novel drug-action mechanisms, (4) applications in intracellular and extracellular compartments and (5) innovative approaches for biochemical assays. Ventana

Medical Systems has just begun publicizing their QDs Map family of immunohistochemistry reagent kits for automated slide processing and fluorescent detection of fixed specimens (www.ventanadiscovery.com). The increased commercial offering of QDs products reflects the desirability of QDs photophysical properties, namely, photostability, single source excitation, narrow emission, multiplexing capabilities, and high quantum yield.

Unfortunately, the lack of reliable and reproducible techniques to conjugate a variety of biomolecules such as antibodies, protein markers, DNA, and RNA to QDs in a methodical way with control over their ratio, orientation, and avidity remains to hinder their ongoing use in clinical diagnostics [3]. In the future, it is to be expected that more commercial products integrating QDs for clinical, diagnostic, and research purposes will be released for public use and manipulation, which will likely give rise to more reliable conjugation techniques as they are further investigated. The outcomes of current research combining nanotechnology and reproductive biology clearly indicate nanoparticles as promising tools for both basic and applied research in animal reproduction [79–84].

Author details

Sapna Jain[1], Seong B. Park[2], Shreekmar R. Pillai[1], Peter L. Ryan[2], Scott T. Willard[2] and Jean M. Feugang[2*]

*Address all correspondence to: jn181@ads.msstate.edu

1 Alabama State University, Alabama, USA

2 Mississippi State University, Mississippi, USA

References

[1] Maiti A, Bhattacharyya S. Quantum dots and applications in medical science. International Journal of Chemical Science and Chemical Engineering. 2013;**3**(2):37-42

[2] Wu M, Wang X, Wang K, Guo Z. An ultrasensitive fluorescent nanosensor for trypsin based on upconversion nanoparticles. Talanta. 2017;**174**(supplement C):797-802

[3] Valizadeh A, Mikaeili H, Samiei M, Farkhani SM, Zarghami N, Kouhi M, Akbarzadeh A, Davaran S. Quantum dots: Synthesis, bioapplications and toxicity. Nanoscale Research Letters. 2012;**7**(1):480

[4] Liu Y, Bose S, Fan W. Effect of size and shape on electronic and optical properties of CdSe quantum dots. Optik—International Journal for Light and Electron Optics. 2018;**155**(supplement C):242-250

[5] Xu R, Huang B, Wang T, Yuan Y, Zhang L, Lu C, Cui Y, Zhang J. Bright and high-photostable inner-Mn-doped core/giant-shell quantum dots. Superlattices and Microstructures. 2017;**111**(supplement C):665-670

[6] Borah P, Siboh D, Kalita PK, Sarma JK, Nath NM. Quantum confinement induced shift in energy band edges and band gap of a spherical quantum dot. Physica B: Condensed Matter. arXiv preprint arXiv:1705.10343 (2017)

[7] Zhang H, Gao X, Liu S, Su X. One-pot synthesis of stable water soluble Mn: ZnSe/ZnS core/shell quantum dots. Journal of Nanoparticle Research. 2013;**15**(6):1749

[8] Wang J, Li Q, Zhou J, Wang Y, Yu L, Peng H, Zhu J. Synthesis, characterization and cells and tissues imaging of carbon quantum dots. Optical Materials. 2017;**72**(supplement C):15-19

[9] Bodas D, Khan-Malek C. Direct patterning of quantum dots on structured PDMS surface. Sensors and Actuators B: Chemical. 2007;**128**(1):168-172

[10] Talapin DV, Rogach AL, Kornowski A, Haase M, Weller H. Highly luminescent monodisperse CdSe and CdSe/ZnS nanocrystals synthesized in a hexadecylamine–trioctylphosphine oxide–trioctylphosphine mixture. Nano Letters. 2001;**1**(4):207-211

[11] Peng X, Schlamp MC, Kadavanich AV, Alivisatos AP. Epitaxial growth of highly luminescent CdSe/CdS core/shell nanocrystals with photostability and electronic accessibility. Journal of the American Chemical Society. 1997;**119**(30):7019-7029

[12] Medintz IL, Mattoussi H, Clapp AR. Potential clinical applications of quantum dots. International Journal of Nanomedicine. 2008;**3**(2):151

[13] Jorge P, Martins MA, Trindade T, Santos JL, Farahi F. Optical fiber sensing using quantum dots. Sensors. 2007;**7**(12):3489-3534

[14] Wang H-Q, Zhang H-L, Li X-Q, Wang J-H, Huang Z-L, Zhao Y-D. Solubilization and bioconjugation of QDs and their application in cell imaging. Journal of Biomedical Materials Research Part A. 2008;**86**(3):833-841

[15] Hao Y, Gan Z, Xu J, Wu X, Chu PK. Poly(ethylene glycol)/carbon quantum dot composite solid films exhibiting intense and tunable blue–red emission. Applied Surface Science. 2014;**311**(Supplement C):490-497

[16] Zhang C, Kuai Y, Cheng H, Liu X, Ma L. Covalent bonding of grafted polymer brushes of poly(poly(ethylene glycol) monomethacrylate) on surface of silicon quantum dots and the activation of the end hydroxyls. Arabian Journal of Chemistry. 2017

[17] Liu W, He Z, Liang J, Zhu Y, Xu H, Yang X. Preparation and characterization of novel fluorescent nanocomposite particles: CdSe/ZnS core-shell quantum dots loaded solid lipid nanoparticles. Journal of Biomedical Materials Research. 2008;**84**(4):1018-1025

[18] Koole R, van Schooneveld MM, Hilhorst J, de Mello DonegÃ¡ C, Hart DCÊ, van Blaaderen A, Vanmaekelbergh D, Meijerink A. On the incorporation mechanism of hydrophobic quantum dots in silica spheres by a reverse microemulsion method. Chemistry of Materials. 2008;**20**(7):2503-2512

[19] Zhu M-Q, Chang E, Sun J, Drezek RA. Surface modification and functionalization of semiconductor quantum dots through reactive coating of silanes in toluene. Journal of Materials Chemistry. 2007;**17**(8):800-805

[20] Wu X, Liu H, Liu J, Haley KN, Treadway JA, Larson JP, Ge N, Peale F, Bruchez MP. Immunofluorescent labeling of cancer marker Her2 and other cellular targets with semiconductor quantum dots. Nature Biotechnology. 2003;**21**(1):41-46

[21] Zhang Y, Clapp A. Overview of stabilizing ligands for biocompatible quantum dot nanocrystals. Sensors. 2011;**11**(12):11036-11055

[22] Medintz IL, Uyeda HT, Goldman ER, Mattoussi H. Quantum dot bioconjugates for imaging, labelling and sensing. Nature Materials. 2005;**4**(6):435-446

[23] Michalet X, Pinaud FF, Bentolila LA, Tsay JM, Doose S, Li JJ, Sundaresan G, AM W, Gambhir SS, Weiss S. Quantum dots for live cells, *in vivo* imaging, and diagnostics. Science. 2005;**307**(5709):538-544

[24] Gerion D. Fluorescence imaging in biology using nanoprobes. Nanotechnologies for the Life Sciences. 2006

[25] Derfus AM, Chan WCW, Bhatia SN. Probing the cytotoxicity of semiconductor quantum dots. Nano Letters. 2004;**4**(1):11-18

[26] Wang L, Nagesha DK, Selvarasah S, Dokmeci MR, Carrier RL. Toxicity of CdSe nanoparticles in Caco-2 cell cultures. Journal of Nanobiotechnology. 2008;**6**(1):11

[27] Tchounwou PB, Yedjou CG, Patlolla AK, Sutton DJ. Heavy metals toxicity and the environment. EXS. 2012;**101**:133-164

[28] Byrne SJ, Williams Y, Davies A, Corr SA, Rakovich A, Gun'ko YK, Rakovich YP, Donegan JF, Volkov Y. "Jelly dots": Synthesis and cytotoxicity studies of CdTe quantum dot-gelatin nanocomposites. Small. 2007;**3**(7):1152-1156

[29] Zhang A, Dong C, Li L, Yin J, Liu H, Huang X, Ren J. Non-blinking (Zn)CuInS/ZnS quantum dots prepared by *In Situ* interfacial alloying approach. Scientific Reports. 2015;**5**:15227

[30] Kirchner C, Liedl T, Kudera S, Pellegrino T, Muñoz Javier A, Gaub HE, Stölzle S, Fertig N, Parak WJ. Cytotoxicity of colloidal CdSe and CdSe/ZnS nanoparticles. Nano Letters. 2005;**5**(2):331-338

[31] Mashinchian O, Johari-Ahar M, Ghaemi B, Rashidi M, Barar J, Omidi Y. Impacts of quantum dots in molecular detection and bioimaging of cancer. BioImpacts: BI. 2014;**4**(3):149

[32] Lee KP, Kelly DP, Schneider PW, Trochimowicz HJ. Inhalation toxicity study on rats exposed to titanium tetrachloride atmospheric hydrolysis products for two years. Toxicology and Applied Pharmacology. 1986;**83**(1):30-45

[33] Hoet PHM, Bruske-Hohlfeld I, Salata OV. Nanoparticles-known and unknown health risks. Journal of Nanobiotechnology. 2004;**2**(1):12

[34] Hoet PHM, Gilissen L, Nemery B. Polyanions protect against the *in vitro* pulmonary toxicity of polycationic paint components associated with the Ardystil syndrome. Toxicology and Applied Pharmacology. 2001;**175**(2):184-190

[35] Hoet PH, Gilissen LP, Leyva M, Nemery B. *In vitro* cytotoxicity of textile paint components linked to the "Ardystil syndrome". Toxicological Sciences: An Official Journal of the Society of Toxicology. 1999;**52**(2):209-216

[36] Su Y, Peng F, Jiang Z, Zhong Y, Lu Y, Jiang X, Huang Q, Fan C, Lee S-T, He Y. *In vivo* distribution, pharmacokinetics, and toxicity of aqueous synthesized cadmium-containing quantum dots. Biomaterials. 2011;**32**(25):5855-5862

[37] Jin S, Hu Y, Gu Z, Liu L, Wu H-C. Application of quantum dots in biological imaging. Journal of Nanomaterials. 2011;**2011**:13

[38] Hoshino A, Hanaki K-I, Suzuki K, Yamamoto K. Applications of T-lymphoma labeled with fluorescent quantum dots to cell tracing markers in mouse body. Biochemical and Biophysical Research Communications. 2004;**314**(1):46-53

[39] Hoshino A, Fujioka K, Oku T, Suga M, Sasaki YF, Ohta T, Yasuhara M, Suzuki K, Yamamoto K. Physicochemical properties and cellular toxicity of nanocrystal quantum dots depend on their surface modification. Nano Letters. 2004;**4**(11):2163-2169

[40] Shiohara A, Hoshino A, Hanaki K, Suzuki K, Yamamoto K. On the cytotoxicity caused by quantum dots. Microbiology and Immunology. 2004;**48**(9):669-675

[41] Lovrić J, Cho SJ, Winnik FM, Maysinger D. Unmodified cadmium telluride quantum dots induce reactive oxygen species formation leading to multiple organelle damage and cell death. Chemistry & Biology. 2005;**12**(11):1227-1234

[42] Dubertret B, Skourides P, Norris DJ, Noireaux V, Brivanlou AH, Libchaber A. *In vivo* imaging of quantum dots encapsulated in phospholipid micelles. Science. 2002;**298**(5599): 1759-1762

[43] Ballou B, Lagerholm BC, Ernst LA, Bruchez MP, Waggoner AS. Noninvasive imaging of quantum dots in mice. Bioconjugate Chemistry. 2004;**15**(1):79-86

[44] Jaiswal JK, Mattoussi H, Mauro JM, Simon SM. Long-term multiple color imaging of live cells using quantum dot bioconjugates. Nature Biotechnology. 2002;**21**:47

[45] Larson DR, Zipfel WR, Williams RM, Clark SW, Bruchez MP, Wise FW, Webb WW. Water-soluble quantum dots for multiphoton fluorescence imaging *in vivo*. Science. 2003;**300**(5624):1434

[46] Voura EB, Jaiswal JK, Mattoussi H, Simon SM. Tracking metastatic tumor cell extravasation with quantum dot nanocrystals and fluorescence emission-scanning microscopy. Nature Medicine. 2004;**10**:993

[47] Hanaki K-I, Momo A, Oku T, Komoto A, Maenosono S, Yamaguchi Y, Yamamoto K. Semiconductor quantum dot/albumin complex is a long-life and highly photostable endosome marker. Biochemical and Biophysical Research Communications. 2003;**302**(3):496-501

[48] Chen F, Gerion D. Fluorescent CdSe/ZnS nanocrystal-peptide conjugates for long-term, nontoxic imaging and nuclear targeting in living cells. Nano Letters. 2004;**4**(10):1827-1832

[49] Chen N, He Y, Su Y, Li X, Huang Q, Wang H, Zhang X, Tai R, Fan C. The cytotoxicity of cadmium-based quantum dots. Biomaterials. 2012;**33**(5):1238-1244

[50] Prasad BR, Nikolskaya N, Connolly D, Smith TJ, Byrne SJ, Gérard VA, Gun'ko YK, Rochev Y. Long-term exposure of CdTe quantum dots on PC12 cellular activity and the determination of optimum non-toxic concentrations for biological use. Journal of Nanobiotechnology. 2010;**8**(1):7

[51] Su Y, He Y, Lu H, Sai L, Li Q, Li W, Wang L, Shen P, Huang Q, Fan C. The cytotoxicity of cadmium based, aqueous phase—Synthesized, quantum dots and its modulation by surface coating. Biomaterials. 2009;**30**(1):19-25

[52] Bruchez M, Moronne M, Gin P, Weiss S, Alivisatos AP. Semiconductor nanocrystals as fluorescent biological labels. Science. 1998;**281**(5385):2013-2016

[53] Tang M, Xing T, Zeng J, Wang H, Li C, Yin S, Yan D, Deng H, Liu J, Wang M, Chen J, Ruan D-Y. Unmodified CdSe quantum dots induce elevation of cytoplasmic calcium levels and impairment of functional properties of sodium channels in rat primary cultured hippocampal neurons. Environmental Health Perspectives. 2008;**116**(7):915-922

[54] Resch-Genger U, Grabolle M, Cavaliere-Jaricot S, Nitschke R, Nann T. Quantum dots *versus* organic dyes as fluorescent labels. Nature Methods. 2008;**5**(9):763-775

[55] Wang Y, Hu R, Lin G, Roy I, Yong K-T. Functionalized quantum dots for biosensing and bioimaging and concerns on toxicity. ACS Applied Materials & Interfaces. 2013;**5**(8):2786-2799

[56] Lu ZS, Li CM. Quantum dot-based nanocomposites for biomedical applications. Current Medicinal Chemistry. 2011;**18**(23):3516-3528

[57] Liu BR, Huang Y-W, Chiang H-J, Lee H-J. Cell-penetrating peptide-functionized quantum dots for intracellular delivery. Journal of Nanoscience and Nanotechnology. 2010;**10**(12):7897-7905

[58] Feugang JM, Youngblood RC, Greene JM, Willard ST, Ryan PL. Self-illuminating quantum dots for non-invasive bioluminescence imaging of mammalian gametes. Journal of Nanobiotechnology. 2015;**13**(1):38

[59] Feugang JM, Youngblood RC, Greene JM, Fahad AS, Monroe WA, Willard ST, Ryan PL. Application of quantum dot nanoparticles for potential non-invasive bio-imaging of mammalian spermatozoa. Journal of Nanobiotechnology. 2012;**10**:45

[60] Wiwanitkit V, Sereemaspun A, Rojanathanes R. Effect of gold nanoparticles on spermatozoa: The first world report. Fertility and Sterility. 2009;**91**(1):e7-e8

[61] Tiedemann D, Taylor U, Rehbock C, Jakobi J, Klein S, Kues WA, Barcikowski S, Rath D. Reprotoxicity of gold, silver, and gold-silver alloy nanoparticles on mammalian gametes. The Analyst. 2014;**139**(5):931-942

[62] Taylor U, Rehbock C, Streich C, Rath D, Barcikowski S. Rational design of gold nanoparticle toxicology assays: A question of exposure scenario, dose and experimental setup. Nanomedicine. 2014;**9**(13):1971-1989

[63] Taylor U, Barchanski A, Petersen S, Kues WA, Baulain U, Gamrad L, Sajti L, Barcikowski S, Rath D. Gold nanoparticles interfere with sperm functionality by membrane adsorption without penetration. Nanotoxicology. 2014;**8**(S1):118-127

[64] Moretti E, Terzuoli G, Renieri T, Iacoponi F, Castellini C, Giordano C, Collodel G. *In vitro* effect of gold and silver nanoparticles on human spermatozoa. Andrologia. 2013;**45**(6):392-396

[65] Amiri G, Valipoor A, Parivar K, Modaresi M, Noori A, Gharamaleki H, Taheri J, Kazemi A. Comparison of toxicity of CdSe: ZnS quantum dots on male reproductive system in different stages of development in mice. International Journal of Fertility & Sterility. 2016;**9**(4):512-520

[66] Chan WCW, Nie S. Quantum dot bioconjugates for ultrasensitive nonisotopic detection. Science. 1998;**281**(5385):2016-2018

[67] Yoo JS, Won N, Kim HB, Bang J, Kim S. *In vivo* imaging of cancer cells with electroporation of quantum dots and multispectral imaging. Journal of Applied Physics. 2010;**107**(12):124702

[68] Liu BR, Li J-F, S-W L, Lee H-J, Huang Y-W, Shannon KB, Aronstam RS. Cellular internalization of quantum dots noncovalently conjugated with arginine-rich cell-penetrating peptides. Journal of Nanoscience and Nanotechnology. 2010;**10**(10):6534-6543

[69] Jeong SH, Kim JH, Yi SM, Lee JP, Kim JH, Sohn KH, Park KL, Kim M-K, Son SW. Assessment of penetration of quantum dots through *in vitro* and *in vivo* human skin using the human skin equivalent model and the tape stripping method. Biochemical and Biophysical Research Communications. 2010;**394**(3):612-615

[70] Si-Han W, YHaC YM. Mesoporous silica nanoparticles as nanocarriers. Chemical Communications. 2011;**47**:9972

[71] Levene MJ, Dombeck DA, Kasischke KA, Molloy RP, Webb WW. *In vivo* multiphoton microscopy of deep brain tissue. Journal of Neurophysiology. 2004;**91**(4):1908-1912

[72] Xu G, Mahajan S, Roy I, Yong K-T. Theranostic quantum dots for crossing blood–brain barrier *in vitro* and providing therapy of HIV-associated encephalopathy. Frontiers in Pharmacology. 2013;**4**

[73] Naji Asfestani M, Rasouli Heikalabad S. A unique structure for the multiplexer in quantum-dot cellular automata to create a revolution in design of nanostructures. Physica B: Condensed Matter. 2017;**512**(supplement C):91-99

[74] Xing Y, Rao J. Quantum dot bioconjugates for *in vitro* diagnostics & *in vivo* imaging. Cancer Biomarkers. 2008;**4**(6):307-319

[75] Pinaud F, Michalet X, Bentolila LA, Tsay JM, Doose S, Li JJ, Iyer G, Weiss S. Advances in fluorescence imaging with quantum dot bio-probes. Biomaterials. 2006;**27**(9):1679-1687

[76] So M-K, Loening AM, Gambhir SS, Rao J. Creating self-illuminating quantum dot conjugates. Natural Protocols. 2006;**1**(3):1160-1164

[77] So MK, Xu C, Loening AM, Gambhir SS, Rao J. Self-illuminating quantum dot conjugates for *in vivo* imaging. Nature Biotechnology. 2006;**24**(3):339-343

[78] Julien A, Park S, Vance C, Ryan P, Willard S, Kouba A, Feugang J. Incorporation and developmental toxicity of quantum dot nanoparticles in amphibian larvae. Reproduction, Fertility and Development. 2017;**29**(1):166-167

[79] Odhiambo JF, DeJarnette JM, Geary TW, Kennedy CE, Suarez SS, Sutovsky M, Sutovsky P. Increased conception rates in beef cattle inseminated with nanopurified bull semen. Biology of Reproduction. 2014;**91**(4):97

[80] Feugang JM. Novel agents for sperm purification, sorting and imaging. Molecular Reproduction and Development. 2017;**9999**:1-10

[81] Durfey CL, Burnett DD, Liao SF, Steadman CS, Crenshaw MA, Clemente HJ, Willard ST, Ryan PL, Feugang JM. Nanotechnology-based selection of boar spermatozoa: Growth development and health assessments of produced offspring. Livestock Science. 2017; **205**(supplement C):137-142

[82] Durfey C, Swistek S, Tan W, Clemente H, Ryan P, Willard S, Feugang J. Beneficial effects of semen purification with magnetic nanoparticles. Mississippi Academy of Science. 2017;**62**(1):164

[83] Durfey C, Liao S, Devost-Burnett D, Dinh T, Crenshaw M, Willard S, Ryan P, Clemente H, Feugang J. Growth and market quality of pigs born from magnetic nanoparticle-treated spermatozoa. Reproduction, Fertility and Development. 2017;**29**(1):141-141

[84] Barkalina N, Jones C, Townley H, Coward K. Functionalization of mesoporous silica nanoparticles with a cell-penetrating peptide to target mammalian sperm *in vitro*. Nanomedicine. 2015;**10**(10):1539-1553

Section 3

Environmental Nanotoxicology

Chapter 7

Bioaccumulation and Toxic Profiling of Nanostructured Particles and Materials

Subas Chandra Dinda

Additional information is available at the end of the chapter

http://dx.doi.org/10.5772/intechopen.74802

Abstract

Use of nanotechnological based formulations and nanomaterials are increasing day-by-day in wide range covering a broad typology of applications, from design and development of targeted drug delivery systems, manufacturing of pesticides, domestic appliances, textiles, to bioremediation engineering. There are therefore concerns about the environmental risks or bioaccumulation-related issues that may arise particularly resulting from the application of drug-loaded nanocarriers or effect of pesticides that reach the natural ecosystems. This is a major threat in the present era and needs to be balanced against their undoubted benefits to human society. The assessment of the physical and chemical properties of nanoparticles and nanomaterials influencing their toxic manifestation due to accumulation in human or in animal organs is still poorly investigated. This chapter reviews the possibilities of bioaccumulation of different nanoscale particles and materials, their potential acute and subacute toxicological profile and their identification and characterization in different organs and tissues of vertebrates.

Keywords: nanomaterials, nanoparticles, bioaccumulation, characterization, toxicity, analytical methods, in vitro studies, in vivo studies

1. Introduction

Nanostructured materials can be easily accessible to the cellular level of a tissue or an organ in comparison to macro scale particles given their small size that makes them prone to passive entrance. On the other hand the functional performance of nanoparticles depends on their size, morphology and chemical nature of its surface which influence the triggering of phagocytosis through the cellular membranes. Moreover, large pay loads may conduce to stress conditions, and thereby increase inflammation and reduce defense action against pathogens

[1, 2]. It is also known that nanoparticles made up of non-degradable or slowly degradable materials may accumulate at the cellular level hindrance the enzymatic activity of functional proteins [3]. The toxic manifestations may also arise due to their chemical composition and structure, surface charge, solubility and aggregation on the site of elimination or filtration such as at the glomerular filtration of nephrons. In this context, it is fundamental to investigate the aggregation and agglomeration of the nanostructured particles and materials, followed in the ionic environment of the biological fluids to elicit its functional activity and resulting toxicity.

Among available manufactured nanomaterials, titanium dioxide (TiO_2) is one of the most widely used to date [4]. Previous studies suggested that TiO_2 nanoparticles production ranged from 7800 to 38,000 metric tons per year in the USA alone and will be approximately 2.5 million metric tons by 2025 [5, 6]. The fate and bioaccumulation of TiO_2 based nanomaterials (nanotubes TiO_2-NTs) and nanoparticles (TiO_2-NPs) in a paddy microcosm over a period of 17 days showed that the Ti levels were the highest in biofilms during the exposure period [7].

Bioaccumulation factors indicated that TiO_2-NPs and TiO_2-NTs were largely transferred from low to superior trophic levels [8]. Considering the potential entries of TiO_2 based nanomaterials in organisms, their bioaccumulation throughout the food chain should be regarded with great concern in terms of the overall health of the natural ecosystems [9].

Aquatic floras play a vital role as primary producers of biomass which constitutes the elementary level of food in trophic chains. Environmental exposure of algae to nanoparticles at potential toxic concentrations may affect natural ecosystems by interfering at the basic energy source of the food webs. As an example, Chandler et al. demonstrated that increased alga mortality occurred upon exposure to nano-ZnO and C60 at the size of 1 ppm in comparison to 10 ppm [10].

At the level of primary consumers it was revealed that bioaccumulation of nanomaterials in Daphnia magna through alga (used as food source), presented a more toxic profile for zinc oxide in comparison to carbon based nanomaterials which may be explained by its higher solubility [10]. In addition Teer and colleagues investigated the effects of nano-Ag derived from coated AgNPs and $AgNO_3$ on phosphate availability given its potential as Ag + ligand and as determinant of phytoplankton productivity [11]. It was observed that both nanparticles accumulated at similar concentrations into the primary producer during high phosphate insult. Yet, $AgNO_3$ accumulation was only increased for low phosphate condition [11].

Regarding studies at superior trophic level, it was shown that goldfish *Carassius auratus* exposed to 10–100 mg/L of TiO_2-NPs significantly accumulated these nanoparticles on tissues and organs, namely: intestine from 42.71 to 110.68 ppb and gills from 4.10 to 9.86 ppb. In addition, overall growth inhibition induced by lipid oxidation in the liver was recorded [12]. In other work, Hanna and colleagues investigated the potential ecological damage of an essential trace element - Cu - derived from CuO engineered nanoparticles (ENPs) exposed marine mussels exposed to 3 mg of CuO ENPs for 4 weeks resulting its clearance of ionic Cu which decreased to 48% leading to a reduction in the invertebrates growth rate in comparison to the control animals, suggesting that CuO ENPs are less toxic than ionic Cu probably due to the slow dissolution rate of the former [13].

Investigation on top predators including humans demonstrated that inhaled nanoscale particles are less cleared by macrophages than large sized particles [14], and are able to translocate to other organs through circulatory or lymphatic drainage which may increase cytokine production and imbalance in redox potential [15] toward oxidation, leading to inflammation or cell death [16]. Nano sized particles and materials were also demonstrated to be taken up by mitochondria and nucleus of the cell [17] causing DNA mutation [18].

To surmount the health risk and potential toxic manifestations or disease associated conditions resulting from nano-devices or nano-structured materials exposure, efficient multifunctional designs are required to make the most of their interesting features while avoid adverse effects.

2. Mechanism of nanoparticle bioaccumulation and toxic action

Metallic elements such as Ni, Cu, Ti and Ag are increasingly considered as nanoparticles components that are being applied to biocides and antimicrobials [19]. Once release to the natural environment their safety profile is of concern. Studies on their toxicological potential were greatly investigated in aquatic species from invertebrates to vertebrates given that these ecosystems compose the ultimate sink of their fate: zooplankton [20, 21], fish [22, 23], algae [24], marine [25] and fresh water crabs [26]. Overall, these works point to three principal factors as potentially responsible for the metallic nanoparticles toxicity, namely: (i) dissolution rate, (ii) cellular uptake and (iii) induced level of oxidative stress and subsequent cellular damage [27].

Nonmetallic carbon-based nanomaterials such as fullerenes and carbon nanotubes, which are under exploration in cancer drug delivery, were reported to be cytotoxic [28]. Obtained data on cell proliferation of human lung cancer cells and human keratinocytes upon exposure to carbon-based nanomaterials revealed decreased cellular viability [29]. Intravenous administration of single-walled carbon nanotubes on mice showed their long term accumulation in the mammalian organs such as liver, lungs and spleen, which was observed using Raman spectroscopy and TEM technique [30]. In addition, decreased levels glutathione (GSH), and increased malondialdehyde (MDA) levels suggested that the toxicity of the carbon nanotubes is due to oxidative stress [31].

Increased levels of reactive oxygen species (ROS) and malondialdehyde (MDA) were further reported after administration of silica nanoparticles (15–46 nm) at a dose of 10–100 μg/mL in human bronchoalveolar carcinoma cells [32].

2.1. Toxicity of nanoparticles due to dissolution

The toxicity of metallic nanoparticles was found to be dependent on the rate of dissolution of the metal ions from its respective nanoparticulate formulations. Bondarenko and colleagues demonstrated that Zn-ions released from ZnO NPs did not exerted significantly different toxic effects in vivo on different aquatic models, neither in vitro on mammalian cells [33].

Differences were only found on its dissolution rate, which were revealed to be non-species dependent [33]. Under aerobic conditions it was also found a positive correlation among the dissolution rate of AgNPs and their toxicological profile upon exposure of bacteria and zooplankton models [34, 35]. In addition, sub-toxic effects of CuO NPs observed in bacteria seem to be associated with dissolved Cu ions liberated from the respective nanoformulation, triggering ROS production and causing cell death [36].

Regarding non-metallic nanoscale particles, biodegradable polymeric nanoparticles made up of poly-(D), L-lactide-co-glycolide) i.e. PLGA and its derivatives it is known that their drug release is based on the dissolution rate of its surface coating polymeric material PLGA [37]. Nevertheless, it was demonstrated in macrophages that the surface coating of these polymeric nanoparticles may be linked to induced toxicity by increasing oxidative stress levels [38].

2.2. Uptake of nanoparticles and their effects on biological membranes

Metallic nano-sized particles as ZnO NPs, with size inferior to 10 nm, were demonstrated to be highly internalized by prokaryotes in vitro causing cellular damage to both Gram negative *Escherichia coli* and Gram positive *Staphylococcus aureus* [39, 40]. This bactericidal potential was also observed for Ag NPs in the size range of 1–10 nm which was shown to be size-dependent [41, 42]. TEM microscopy confirmed bacteria internalization and showed a uniform pattern of distribution [40]. Even at non cytotoxic concentrations cellular uptake occurred [43].

Regarding the mechanisms of internalization, it was proposed that non-specific diffusion and membrane damage, and porins-specific intake are the potential modes through which the NPs could transpose bacterial wall [44]. Cellular uptake depends on multiple intrinsic (e.g. specific and dependent on the types of cell, tissue or organs) and extrinsic (e.g. NPs size and coating) variables [45]. ZnO NPs functionalization was demonstrated to affect bacteria membrane permeability increasing cellular uptake levels [46]. The hydroxyl groups of polyvinylalcohol (PVA) macromolecule existent on the surface of the coated ZnO NPs was found to disrupt the cellular membrane given its alkaline nature [46]. In case of Ag NPs, it has been shown that certain sizes of these metallic nanoparticles attached to Gram-negative *Escherichia coli* wall, resulting in its perforation which lead to the cell death [47], given that their direct contact with bacteria facilitated their dissolution at cell-NP interface and thus, enhanced their antibacterial effects [48].

In this regard, cationic-NPs have been reported as potential nanoparticles to cause pronounced disruption of plasma membrane integrity leading to mitochondrial and lysosomal damage, and production of autophagosomes, more than anionic-NPs [49]. Also, nonphagocytic cells were demonstrated to ingest cationic NPs to higher extent than anionic NPs [50]. Taking into account their surface charge nanoparticles can influence the selectivity and efficacy of drug delivery and imaging by selecting either a phagocytic or non-phagocytic pathway [51].

To evaluate cellular uptake of polymeric nanoparticles composed of poly-(lactic-co-glycolic acid) (PLGA) and coated with PVA and vitamin E TPGS, human colon adenocarcinoma cells were exposed in vitro [52]. It was found that vitamin E TPGS-coated PLGA NPs showed 1.4 folds higher cellular uptake than that of PVA-coated PLGA nanoparticles [53]. Cryo-SEM and

TEM analysis confirmed the cellular internalization, indicating that NPs surface modification with vitamin E TGPS and PLGA enhance loading of chemotherapeutic agents to be administered through oral route for cancer therapy [54].

2.3. Oxidative stress induced via nanoparticle and nanomaterial cellular uptake

Oxidative damage, caused in consequence of cellular internalization of NPs, has been considered one of the main causes of NPs cytotoxicity [55]. ZnO, CuO and Ag NPs, were already reported as ROS levels disruptors in aquatic microorganisms [56].

Also, the toxic effects of ROS generating potential of engineered Ag NPs were tested on different strains of recombinant *Escherichia coli* mutants and that of wild type strain [57]. The results showed that mutant strains of recombinant type were 15 folds more responsive to Ag NPs than the wild type [58]; although, analogous effects were observed for Ag ions [59]. In eukaryotes, induction of ROS by Ag NPs exposure was also observed in zebrafish Danio rerio [60].

It is well known that ions of redox-active metals, including Cu, may yield free radicals via the Fenton-type reaction and inflict intracellular oxidative stress [61, 62]. The Cu (II) ions can be transformed to Cu (I) ions in the presence of biological reducing agents such as ascorbic acid or glutathione (GSH), and consequently generate reactive hydroxyl radicals from hydrogen peroxide [63]. Using luminescent bacterial tests in recombinant *Escherichia coli* strains, ROS generating potential of aqueous suspensions of CuO NPs was demonstrated [58]. It has also been shown that CuO NPs induced oxidative stress and DNA damage in these recombinant bacteria strains at low subtoxic concentrations of 0.1 mg Cu/L [58].

In vivo investigation of the exposure effects of ZnO NPs on isolated rat liver mitochondria showed increased mitochondrial membrane permeability prone to energy dissipation [64], ROS levels production caused impairment of the mitochondrial respiratory chain and even apoptosis [65]. Similarly, the experiments on isolated human liver cells in-vitro showed ROS triggered mitochondrial pathway resulting apoptosis and cell death [65].

Moreover, it was also demonstrated a concentration dependent genotoxicity derived from exposure of several nanoparticles composed of α-alumina, β-alumina, SiO_2, SbO_5, and Fe_2O_3 through generation of excess ROS, which leaded to DNA strand breakage [66]. DNA cleavage, an indicator of irreversible completion of apoptosis, occurred in organisms exposed to 500 μg/kg of the above-mentioned precipitated nanoparticles [67]. Moreover, cleavage of inter-nucleosomal DNA ladder bands occurred upon exposure to 500 μg/kg of γ-alumina and α-alumina [66].

3. Conclusion

Nano-scale particles and materials applications are increasing significantly and thus their potential environmental and human health risks urges consideration. Analytical methods are required to reliably detect and characterize nanoparticles and nanomaterials in the natural environments, and their properties once interacting with complexes matrices such as air, soil and water, as well as food and consumer products. Advances in engineering particles and materials testing, either qualitative or quantitative determinations, should be pursued to ensure a safety profile, including the development of standard guidelines. In addition to the toxicological studies, various uptake paths have to be investigated, including dermal, oral and intestinal routes, as well as bioaccumulation trials and potential long-term effects consideration. Research into new analytical methods is further required to address the special properties for the establishment of ecological and human health risk factors.

Disclosure statement

No conflicts of interest.

Author details

Subas Chandra Dinda

Address all correspondence to: subas.dinda@rediffmail.com

School of Pharmacy, C.H.S., Mekelle University, Mekelle, Ethiopia

References

[1] Krol S, Macrez R, Docagne F, Defer G, Laurent S, Rahman M, Hajipour MJ, Kehoe PG, Mahmoudi M. Therapeutic benefits from nanoparticles: The potential significance of nanoscience in diseases with compromise to the blood brain barrier. Chemical Reviews. 2013;**113**(3):1877-1903

[2] Dykman LA, Khlebtsov NG. Uptake of engineered gold nanoparticles into mammalian cells. Chemical Reviews. 2014;**114**(2):1258-1288

[3] Singh R, Lillard JW Jr. Nanoparticle-based targeted drug delivery. Experimental and Molecular Pathology. 2009;**86**(3):215-223

[4] Shi H, Magaye R, Castranova V, Zhao J. Titanium dioxide nanoparticles: A review of current toxicological data. Particle and Fibre Toxicology. 2013;**10**:15. DOI: 10.1186/1743-8977-10-15

[5] Hendren CO, Mesnard X, Dröge J, Wiesner MR. Estimating production data for five engineered nanomaterials as a basis for exposure assessment. Environmental Science & Technology. 2011;**45**(7):2562-2569

[6] Robichaud CO, Uyar AE, Darby MR, Zucker LG, Wiesner MR. Estimates of upper bounds and trends in nano-TiO_2 production as a basis for exposure assessment. Environmental Science & Technology. 2009;**43**(12):4227-4233

[7] Kim JI, Park HG, Chang KH, Nam DH, Yeo MK. Trophic transfer of nano-TiO2 in a paddy microcosm: A comparison of single-dose versus sequential multi-dose exposures. Environmental Pollution. 2016;**212**:316-324. DOI: 10.1016/j.envpol.2016.01.076

[8] Borgå K, Fisk AT, Hargrave B, Hoekstra PF, Swackhamer D, Muir DCG. Bioaccumulation factors for PCBs revisited. Environmental Science & Technology. 2005;**39**:4523-4532

[9] Yeo M-K, Nam D-H. Influence of different types of nanomaterials on their bioaccumulation in a paddy microcosm: A comparison of TiO_2 nanoparticles and nanotubes. Environmental Pollution. 2013;**178**:166-172

[10] Chandler JL. Toxicity and bioaccumulation of nanomaterial in aquatic species. Journal of the U.S. SJWP. 2007. DOI: 10.2175/SJWP(2007)1:01

[11] McTeer J, Dean AP, White KN, Pittman JK. Bioaccumulation of silver nanoparticles into Daphnia magna from a fresh water algal diet and the impact of phosphate availability. Journal Nanotoxicology. 2014;**8**(3):305-316. DOI: 10.3109/17435390.2013.778346

[12] Ates M, Demir V, Adiguzel R, Arslan Z. Bioaccumulation, subacute toxicity, and tissue distribution of engineered titanium dioxide nanoparticles in goldfish (*Carassius auratus*). Journal of Nanomaterials. 2013; Article ID 460518:1-6. DOI: 10.1155/2013/460518

[13] Hanna SK, Miller RJ, Lenihan HS. Accumulation and toxicity of copper oxide engineered nanoparticles in a marine mussel. Nanomaterials. 2014;**4**:535-547. DOI: 10.3390/nano4030535

[14] Zhu M-T, Feng W-Y, Wang Y, Wang B, Wang M, Ouyang H, Zhao Y-L, Chai Z-F. Particokinetics and extrapulmonary translocation of Intratracheally instilled ferric oxide nanoparticles in rats and the potential health risk assessment. Toxicological Sciences. 2009;**107**(2):342-351

[15] Hussain T, Tan B, Yin Y, Blachier F, Tossou MCB, Rahu N. Oxidative stress and inflammation: What polyphenols can do for us? Oxidative Medicine and Cellular Longevity. 2016;**2016**:1-9 http://dx.doi.org/10.1155/2016/7432797

[16] Oberdörster G et al. Nanotoxicology: An emerging discipline evolving from studies of ultrafine particles. Environmental Health Perspectives. 2005;**113**(7):823-839. PMC 1257642 . PMID 16002369. DOI: 10.1289/ehp.7339

[17] Li N, Sioutas C, Cho A, et al. Ultrafine particulate pollutants induce oxidative stress and mitochondrial damage. Environmental Health Perspectives. 2003;**111**(4):455-460. PMC 1241427. PMID 12676598. DOI: 10.1289/ehp.6000

[18] Geiser M et al. Ultrafine particles cross cellular membranes by nonphagocytic mechanisms in lungs and in cultured cells. Environmental Health Perspectives. 2005;**113**(11):1555-1560. PMC 1310918. PMID 16263511. DOI: 10.1289/ehp.8006

[19] Raghunath A, Perumal E. Metal oxide nanoparticles as antimicrobial agents: A promise for the future. International Journal of Antimicrobial Agents. 2017;**49**(2):37-152. DOI: 10.1016/j.ijantimicag.2016.11.011

[20] Park S-Y, Choi J. Geno- and ecotoxicity evaluation of silver nanoparticles in fresh water crustacean *Daphnia magna*. Environmental Engineering Research. 2010;**15**(1):23-27. DOI: 10.4491/eer.2010.15.1.428

[21] Heinlaan M, Kahru A, Kasemets K, Arbeille B, Prensier G, Dubourguier H-C. Changes in the *Daphnia magna* midgut upon ingestion of copper oxide nanoparticles: A transmission electron microscopy study. Water Research. 2011;**45**:179-190. DOI: 10.1016/j.watres.2010.08.026

[22] Scown TM, Santaos EM, Johnston BD, GAiser B, Baalousha M, Mitov S, Lead JR, Stone V, Fernandes TF, Jepson M, van Aerle R, Tyler CR. Effects of aqueous exposure to silver nanoparticles of different sizes in rainbow trout. Toxicological Sciences. 2010;**115**(2):521-534. DOI: 10.1093/toxsci/kfq076

[23] AshaRani PV, Wu YL, Gong Z, Valiyaveettil S. Toxicity of silver nanoparticles in zebrafish models. Nanotechnology. 2008;**19**(25):279-290. DOI: 10.1088/0957-4484/19/255102

[24] Oukarroum A, Bras S, Perreault F, Popovic R. Inhibitory effects of silver nanoparticles in two green algae, *Chlorella vulgaris* and *Dunaliella tertiolecta*. Ecotoxicology and Environmental Safety. 2012;**78**(1):80-85. DOI: 10.1016/j.ecoenv.2011.11.012

[25] Pan L, Zhang H. Metallothionein, antioxidant enzymes and DNA strand breaks as biomarkers of Cd exposure in a marine crab, *Charybdis japonica*. Comparative Biochemistry and Physiology, Part C. 2006;**144**(1):67-75. DOI: 10.1016/j.cbpc.2006.06.001

[26] Walters CR, Cheng P, Pool E, Somerset V. Effect of temperature on oxidative stress parameters and enzyme activity in the tissues of cape river crab (Potamonautes perlatus) following exposure to silver nanoparticles (AgNP). Journal of Toxicology and Environmental Health, Part A. 2016;**79**(2):1-11. DOI: 10.1080/15287394.2015.1106357

[27] Ivask A, Juganson K, Bondarenko O, Mortimer M, Aruoja V, Kasemets K, Blinova I, Heinlaan M, Slaveykova V, Kahru A. Mechanisms of toxic action of Ag, ZnO and CuO nanoparticles to selected ecotoxicological test organisms and mammalian cells in vitro: A comparative review. Nanotoxicology. 2014;**8**(S1):57-71. DOI: 10.3109/17435390.2013.855831

[28] Magrez A, Kasas S, Salicio V, Pasquier N, Seo JW, Celio M, Catsicas S, Schwaller B, Forro L. Cellular toxicity of carbon-based nanomaterials. Nano Letters. 2006;**6**(6):1121-1125

[29] Herzog E, Casey A, Lyng FM, Chambers G, Byrne HJ, Davoren M. A new approach to the toxicity testing of carbon-based nanomaterials the clonogenic assay. Toxicology Letters. 2007;**174**(1-3):49-60

[30] Principi E, Girardello R, Bruno A, Manni I, Gini E, Pagani A, Grimaldi A, Ivaldi F, Congiu T, De Stefano D, Piaggio G, de Eguileor M, Noonan DM, Albini A. Systemic distribution of single-walled carbon nanotubes in a novel model: Alteration of biochemical parameters, metabolic functions, liver accumulation, and inflammation in vivo. International Journal of Nanomedicine. 2016;**11**:4299-4316. DOI: 10.2147/IJN.S109950

[31] Yang ST, Wang X, Jia G, Gu Y, Wang T, Nie H, Ge C, Wang H, Liu Y. Long-term accumulation and low toxicity of single-walled carbon nanotubes in intravenously exposed mice. Toxicology Letters. 2008;**181**(7):182-189

[32] Lin W, Huang YW, Zhou XD, Ma Y. In vitro toxicity of silica nanoparticles in human lung cancer cells. Toxicology and Applied Pharmacology. 2006;**217**(3):252-259

[33] Bondarenko O, Ivask A, Ka kinen A, Kurvet I, Kahru A. Particle cell contact enhances antibacterial activity of silver nanoparticles. PLoS One. 2013;**8**:e64060

[34] Boenigk J, Beisser D, Zimmermann S, Bock C, Jakobi J, Grabner D, Großmann L, Rahmann S, Barcikowski S, Sures B. Effects of silver nitrate and silver nanoparticles on a planktonic community: General trends after short-term exposure. PLoS One. 2014;**9**(4):e95340. DOI: 10.1371/journal.pone.0095340

[35] Hoheisel SM, Diamond S, Mount D. Comparison of nanosilver and ionic silver toxicity in *Daphnia magna* and *Pimephales promelas*. Environmental Toxicology and Chemistry. 2012;**31**:2557-2563

[36] Bondarenko O, Ivask A, Kakinen A, Kahru A. Sub-toxic effects of CuO nanoparticles on bacteria: Kinetics, role of Cu ions and possible mechanisms of action. Environmental Pollution. 2012;**169**:81-89

[37] Panyam J, Labhasetwar V. Biodegradable nano-particles for drug and gene delivery to cells and tissue. Advanced Drug Delivery Reviews. 2003;**55**(3):329-347

[38] Grabowski N, Hillaireau H, Vergnaud J, Tsapis N, Pallardy M, Kerdine-Rö me S, Fattal E. Surface coating mediates the toxicity of polymeric nanoparticles towards human-like macrophages. International Journal of Pharmseutics. 2015;**482**(1-2):75-83

[39] Applerot G, Lipovsky A, Dror R, Perkas N, Nitzan Y, Lubart R, Gedanken A. Enhanced antibacterial activity of nanocrystalline ZnO due to increased ROS-mediated cell injury. Advanced Functional Materials. 2009;**19**:842-852

[40] Brayner R, Ferrari-Iliou R, Brivois N, Djediat S, Benedetti MF, Fie vet F. Toxicological impact studies based on Escherichia coli bacteria in ultrafine ZnO nanoparticles colloidal medium. Nano Letters. 2006;**6**:866-870. DOI: 10.1021/nl052326h

[41] Lok C-N, Ho C-M, Chen R, He Q-Y, Yu W-Y, Sun H, et al. Silver nanoparticles: Partial oxidation and antibacterial activities. Journal of Biological Inorganic Chemistry. 2007; **12**:527-534

[42] Morones JR, Elechiguerra JL, Camacho A, Holt K, Kouri JB, Rami rez JT, Yacaman MJ. The bactericidal effect of silver nanoparticles. Nanotechnology. 2005;**16**:2346-2353

[43] Gliga AR, Skoglund S, Wallinder IO, Fadeel B, Karlsson HL. Size-dependent cytotoxicity of silver nanoparticles in human lung cells: The role of cellular uptake, agglomeration and Ag release. Particle and Fibre Toxicology. 2014;**11**:11. DOI: 10.1186/1743-8977-11-11

[44] Kumar A, Pandey AK, Singh SS, Shanker R, Dhawan A. Cellular uptake and mutagenic potential of metal oxide nanoparticles in bacterial cells. Chemosphere. 2011;**83**:1124-1132

[45] Treuel L, Jiang X, Nienhaus GU. New views on cellular uptake and trafficking of manufactured nanoparticles. Journal of the Royal Society Interface. 2013;**10**(82):20120939. DOI: 10.1098/rsif.2012.0939

[46] Huang Z, Zheng X, Yan D, Yin G, Liao X, Kang Y, Yao Y, Huang D, Hao B. Toxicological effect of ZnO nanoparticles based on bacteria. Langmuir. 2008;**24**(8):4140-4144. DOI: 10.1021/la7035949

[47] Aminul HM. Experiment-based quantitative modeling for the antibacterial activity of silver nanoparticles. [Theses and Dissertations] 2017. p. 2445 http://scholarworks.uark.edu/etd/2445

[48] Gogoi SK, Gopinath P, Paul A, Ramesh A, Ghosh SS, Chattopadhyay A. Green fluorescent protein-expressing *Escherichia coli* as a model system for investigating the antimicrobial activities of silver nanoparticles. Langmuir. 2006;**22**:9322-9328

[49] Ruenraroengsak P, Tetley TD. Differential bioreactivity of neutral, cationic and anionic polystyrene nanoparticles with cells from the human alveolar compartment: Robust response of alveolar type 1 epithelial cells. Particle and Fibre Toxicology. 2015;**12**:19. DOI: 10.1186/s12989-015-0091-7

[50] Fleischer CC, Payne CK. Nanoparticle-cell interactions: Molecular structure of the protein corona and cellular outcomes. Accounts of Chemical Research. 2014;**47**(8):2651-2659. DOI: 10.1021/ar500190q

[51] Fröhlich E. The role of surface charge in cellular uptake and cytotoxicity of medical nanoparticles. International Journal of Nanomedicine. 2012;**7**:5577-5591

[52] Feng SS, Zeng W, Teng Lim Y, Zhao L, Yin Win K, Oakley R, Hin Teoh S, Hang Lee RC, Pan S. Vitamin E TPGS-emulsified poly (lactic-co-glycolic acid) nanoparticles for cardiovascular restenosis treatment. Nanomedicine (London, England). 2007;**2**(3):333-344

[53] Win KY, Feng S-S. Effects of particle size and surface coating on cellular uptake of polymeric nanoparticles for oral delivery of anticancer drugs. Biomaterials. 2005;**26**(15): 2713-2722

[54] Win KY, Feng SS. In vitro and in vivo studies on vitamin E TPGS emulsified poly (D,L-lactic-co-glycolic acid) nanoparticles for paclitaxel formulation. Biomaterials. 2006;**27**(10): 2285-2222

[55] Blokhina O, Virolainen E, Fagerstedt KV. Antioxidants, oxidative damage and oxygen deprivation stress: A review. Annals of Botany (London). 2003;**91**:179-194

[56] Von Moos N, Slaveykova VI. Oxidative stress induced by inorganic nanoparticles in bacteria and aquatic microalgae – State of the art and knowledge gaps. Nanotoxicology. 2014;**8**(6):605-630

[57] Jensen KF. The *Escherichia coli* K-12 "wild types" W3110 and MG1655 have an rph frame-shift mutation that leads to pyrimidine starvation due to low pyrE expression levels. Journal of Bacteriology. 1993;**175**(11):3401-3407. DOI: 10.1128/jb.175.11.3401-3407.1993

[58] Ivask A, Bondarenko O, Jepihhina N, Kahru A. Profiling of the reactive oxygen species-related ecotoxicity of CuO, ZnO, TiO_2, silver, and fullerene nanoparticles using a set of recombinant luminescent *Escherichia coli* strains: Differentiating the impact of particles and solubilised metals. Analytical and Bioanalytical Chemistry. 2010;**398**:701-716

[59] Ivask A, ElBadawy A, Kaweeteerawat C, Boren D, Fischer H, Zhaoxia CHC, Liu R, Tolaymat T, Telesca D, Zink JI, Cohen Y, Holden PA, Godwin HA. Toxicity mechanisms in *Escherichia coli* vary for silver nanoparticles and differ from ionic silver. ACS Nano. 2014;**8**(1):374-386. DOI: 10.1021/nn4044047

[60] Choi O, Hu Z. Size dependent and reactive oxygen species related nanosilver toxicity to nitrifying bacteria. Environmental Science & Technology. 2008;**42**:4583-4588

[61] Valko M, Morris H, Cronin MTD. Metals, toxicity and oxidative stress. Current Medicinal Chemistry. 2005;**12**:1161-1208

[62] Thomas C, Mackey MM, Diaz AA, Cox DP. Hydroxyl radical is produced via the fenton reaction in submitochondrial particles under oxidative stress: Implications for diseases associated with iron accumulation. Redox Report. 2009;**14**(3):102-108

[63] Vidrio E, Jung H, Anastasio C. Generation of hydroxyl radicals from dissolved transition metals in surrogate lung fluid solutions. Atmospheric Environment. 2008;**42**(18):4369-4379

[64] Li J-h, Liu X-r, Zhang Y, Tian F-f, Zhao G-y, Yu Q-l-y, Jiang F-l, Liu Y. Toxicity of nano zinc oxide to mitochondria. Toxicology Research. 2012;**1**:137-144. DOI: 10.1039/C2TX20016C

[65] Arora S, Jain J, Rajwade JM, Paknikar KM. Cellular responses induced by silver nanoparticles: In vitro studies. Toxicology Letters. 2008;**179**:93-100

[66] Oberholster PJ, Musee N, Noth A-M, Chelule PK, Focke WW, Ashton PJ. Assessment of the effect of nanomaterials on sediment-dwelling invertebrate Chironomus tentans larvae. Ecotoxicology and Environmental Safety. 2011;**74**:416-423. DOI: 10.1016/j.ecoenv.2010.12.012

[67] Lim HK, Asharani PV, Prakash Hande M. Enhanced genotoxicity of silver nanoparticles in DNA repair deficient mammalian cells. Frontiers in Genetics. 2012;**3**:104. DOI: 10.3389/fgene.2012.001049